THE CULT OF
STATISTICAL
SIGNIFICANCE

*How the Standard Error
Costs Us Jobs,
Justice, and Lives*

By Stephen T. Ziliak

and

Deirdre N. McCloskey

The University of Michigan Press ～ *Ann Arbor*

Copyright © by the University of Michigan 2008
All rights reserved
Published in the United States of America by
The University of Michigan Press
Manufactured in the United States of America
⊗ Printed on acid-free paper

2019   2018   2017   2016   13   12   11   10

A CIP catalog record for this book is available from the British Library.

Library of Congress Cataloging-in-Publication Data

Ziliak, Stephen Thomas, 1963–
    The cult of statistical significance : how the standard error costs us jobs,
justice, and lives / by Stephen T. Ziliak and Deirdre N. McCloskey.
        p. cm. — (Economics, cognition, and society series)
    Includes index.
    ISBN-13: 978-0-472-07007-7 (cloth : alk. paper)
    ISBN-10: 0-472-07007-X (cloth : alk. paper)
    ISBN-13: 978-0-472-05007-9 (pbk. : alk. paper)
    ISBN-10: 0-472-05007-9 (pbk. : alk. paper)
        1. Economics—Statistical methods.   2. Statistics—Social aspects.   3. Statistical
hypothesis testing—Social aspects.   I. McCloskey, Deirdre N.   II. Title.

HB137.Z55   2007
330.01'5195—dc22   2007035401

The Cult of Statistical Significance

ECONOMICS
COGNITION
AND SOCIETY

This series provides a forum for theoretical and empirical investigations of social phenomena. It promotes works that focus on the interactions among cognitive processes, individual behavior, and social outcomes. It is especially open to interdisciplinary books that are genuinely integrative.

Editor: Timur Kuran
Editorial Board: Tyler Cowen   Avner Greif
                        Diego Gambetta   Viktor Vanberg

Titles in the Series

Stephen T. Ziliak and Deirdre N. McCloskey. *The Cult of Statistical Significance: How the Standard Error Costs Us Jobs, Justice, and Lives*

Eirik G. Furubotn and Rudolf Richter. *Institutions and Economic Theory: The Contribution of the New Institutional Economics,* Second Edition

Tyler Cowen. *Markets and Cultural Voices: Liberty vs. Power in the Lives of Mexican Amate Painters*

Thráinn Eggertsson. *Imperfect Institutions: Possibilities and Limits of Reform*

Vernon W. Ruttan. *Social Science Knowledge and Economic Development: An Institutional Design Perspective*

Phillip J. Nelson and Kenneth V. Greene. *Signaling Goodness: Social Rules and Public Choice*

Stephen Knack, Editor. *Democracy, Governance, and Growth*

Omar Azfar and Charles A. Cadwell, Editors. *Market-Augmenting Government: The Institutional Foundations for Prosperity*

Randall G. Holcombe. *From Liberty to Democracy: The Transformation of American Government*

David T. Beito, Peter Gordon, and Alexander Tabarrok, Editors. *The Voluntary City: Choice, Community, and Civil Society*

Alexander J. Field. *Altruistically Inclined? The Behavioral Sciences, Evolutionary Theory, and the Origins of Reciprocity*

David George. *Preference Pollution: How Markets Create the Desires We Dislike*

Julian L. Simon. *The Great Breakthrough and Its Cause*

E. L. Jones. *Growth Recurring: Economic Change in World History*

Rosemary L. Hopcroft. *Regions, Institutions, and Agrarian Change in European History*

Lee J. Alston, Gary D. Libecap, and Bernardo Mueller. *Titles, Conflict, and Land Use: The Development of Property Rights and Land Reform on the Brazilian Amazon Frontier*

Daniel B. Klein, Editor. *Reputation: Studies in the Voluntary Elicitation of Good Conduct*

Richard A. Easterlin. *Growth Triumphant: The Twenty-first Century in Historical Perspective*

(continues on last page)

To Lawrence *and* Barbara Ziliak,
the older generation

❧

To Connor *and* Lily McCloskey,
the younger

❧

And to the memory of
William H. Kruskal
(1919–2005)

The History of Science has suffered greatly from the use by teachers of second-hand material, and the consequent obliteration of the circumstances and the intellectual atmosphere in which the great discoveries of the past were made. A first-hand study is always instructive, and often . . . full of surprises.

RONALD A. FISHER, 1955

In 1908 "Student," William Sealy Gosset (1876–1937), invented a statistical instrument that would change the life and social sciences. Now those sciences are being ruined by it in a way that Student himself always warned it could. (Photo courtesy of the Galton Laboratory, University College London, and *Annals of Human Genetics.*)

Contents

A Significant Problem   1

In many of the life and human sciences the existence/whether question of the philosophical disciplines has substituted for the size matters/how much question of the scientific disciplines. The substitution is causing a loss of jobs, justice, profits, environmental quality, and even life. The substitution we are worrying about here is called "statistical significance"—a qualitative, philosophical rule that has substituted for a quantitative, scientific magnitude and judgment.

1. Dieting "Significance" and the Case of Vioxx   23

Since R. A. Fisher (1890–1962) the sciences that have put statistical significance at their centers have misused it. They have lost interest in estimating and testing for the actual effects of drugs or fertilizers or economic policies. The big problem began when Fisher ignored the size-matters/how-much question central to a statistical test invented by William Sealy Gosset (1876–1937), so-called Student's *t*. Fisher substituted for it a qualitative question concerning the "existence" of an effect, by which he meant "low sampling error by an arbitrary standard of variance." Forgetting after Fisher what is known in statistics as a "minimax strategy," or other "loss function," many sciences have fallen into a sizeless stare. They seek sampling precision only. And they end by asserting that sampling precision just *is* oomph, magnitude, practical significance. The minke and sperm whales of Antarctica and the users and makers of Vioxx are some of the recent victims of this bizarre ritual.

2. The Sizeless Stare of Statistical Significance   33

Crossing frantically a busy street to save your child from certain death is a good gamble. Crossing frantically to get another mustard packet for your hot dog is not. The size of the potential loss if you don't hurry to save your child is larger, most will agree, than the potential loss if you don't get the mustard. But a majority of scientists in economics, medicine, and other statistical fields appear not to grasp the difference. If they have been trained in exclusively Fisherian methods (and nearly all of them have) they look only for a probability of success in the crossing—the existence of a probability of success better than .99 or .95 or .90, and this within the restricted frame of sampling—ignoring in any spiritual or financial currency the value of the prize and the expected cost of pursuing it. In the life and human sciences a majority of scientists look at the world with what we have dubbed "the sizeless stare of statistical significance."

3. What the Sizeless Scientists Say in Defense 42

The sizeless scientists act as if they believe the *size* of an effect does not matter. In their hearts they do care about size, magnitude, oomph. But strangely they don't measure it. They substitute "significance" measured in Fisher's way. Then they take the substitution a step further by limiting their concern for error to errors in sampling only. And then they take it a step further still, reducing all errors in sampling to one kind of error—that of excessive skepticism, "Type I error." Their main line of defense for this surprising and unscientific procedure is that, after all, "*statistical* significance," which they have calculated, is "objective." But so too are the digits in the New York City telephone directory, objective, and the spins of a roulette wheel. These are no more relevant to the task of finding out the sizes and properties of viruses or star clusters or investment rates of return than is statistical significance. In short, statistical scientists after Fisher neither test nor estimate, really, truly. They "testimate."

4. Better Practice: β-Importance vs. α-"Significance" 57

The most popular test was invented, we've noted, by Gosset, better known by his pen name "Student," a chemist and brewer at Guinness in Dublin. Gosset didn't think his test was very important to his main goal, which was of course brewing a good beer at a good price. The test, Gosset warned right from the beginning, does *not* deal with substantive importance. It does not begin to measure what Gosset called "real error" and "pecuniary advantage," two terms worth reviving in current statistical practice. But Karl Pearson and especially the amazing Ronald Fisher didn't listen. In two great books written and revised during the 1920s and 1930s, Fisher imposed a Rule of Two: if a result departs from an assumed hypothesis by two or more standard deviations of its own sampling variation, regardless of the size of the prize and the expected cost of going for it, then it is to be called a "significant" scientific finding. If not, not. Fisher told the subjectivity-phobic scientists that if they wanted to raise their studies "to the rank of sciences" they must employ his rule. He later urged them to ignore the size-matters/how-much approaches of Gosset, Neyman, Egon Pearson, Wald, Jeffreys, Deming, Shewhart, and Savage. Most statistical scientists listened to Fisher.

5. A Lot Can Go Wrong in the Use of Significance Tests in Economics 62

We ourselves in our home field of economics were long enchanted by Fisherian significance and the Rule of Two. But at length we came to wonder why the correlation of prices at home with prices abroad must be "within two standard deviations of 1.0 in the sample" before one could speak about the integration of world markets. And we came to think it strange that the U.S. Department of Labor refused to discuss black teenage unemployment rates of 30 or 40 percent because they were, by Fisher's circumscribed definition, "insignificant." After being told repeatedly, if implausibly, that such mistakes in the use of Gosset's test were *not* common in economics, we developed in the 1990s a questionnaire to test in economics articles for economic as against statistical significance. We applied it to the behavior of our tribe during the 1980s.

6. A Lot Did Go Wrong in the *American Economic Review* during the 1980s 74

We did not study the scientific writings of amateurs. On the contrary, we studied the *American Economic Review* (known to its friends as the *AER*), a leading journal of economics. With questionnaire in hand we read every full-length article it published that used a test of statistical significance from January 1980 to December 1989. As we expected, in the 1980s more than 70 percent of the articles made the significant mistake of R. A. Fisher.

7. Is Economic Practice Improving?   79

We published our article in 1996. Some of our colleagues replied, "In the old days [of the 1980s] people made that mistake, but [in the 1990s] we modern sophisticates do not." So in 2004 we published a follow-up study, reading all the articles published in the *AER* in the next decade, the 1990s. Sadly, our colleagues were again mistaken. Since the 1980s the practice in important respects got worse, not better. About 80 percent of the articles made the mistaken Fisherian substitution, failing to examine the magnitudes of their results. And less than 10 percent showed full concern for oomph. In a leading journal of economics, in other words, nine out of ten articles in the 1990s acted as if size doesn't matter for deciding whether a number is big or small, whether an effect is big or small enough to matter. The significance asterisk, the flickering star of *, has become a totem of economic belief.

8. How Big Is Big in Economics?   89

Does globalization hurt the poor, does the minimum wage increase unemployment, does world money cause inflation, does public welfare undermine self-reliance? Such scientific questions are always matters of economic significance. *How much* hurt, increase, cause, undermining? Size matters. Oomph is what we seek. But that is not what is found by the statistical methods of modern economics.

9. What the Sizeless Stare Costs, Economically Speaking   98

Sizeless economic research has produced mistaken findings about purchasing power parity, unemployment programs, monetary policy, rational addiction, and the minimum wage. In truth, it has vitiated most econometric findings since the 1920s and virtually all of them since the significance error was institutionalized in the 1940s. The conclusions of Fisherian studies might occasionally be correct. But only by accident.

10. How Economics Stays That Way: The Textbooks and the Referees   106

Now assistant professors are not to blame. Look rather at the report card of their teachers and editors and referees—notwithstanding cries of anguish from the wise Savages, Zellners, Grangers, and Leamers of the economics profession. Economists received a quiet warning by F. Y. Edgeworth in 1885—too quiet, it seems—that sampling precision is not the same as oomph. They ignored it and have ignored other warnings, too.

11. The Not-Boring Rise of Significance in Psychology   123

Did other fields, such as psychology, do the same? Yes. In 1919 Edwin Boring warned his fellow psychologists about confusing so-called statistical with actual significance. Boring was a famous experimentalist at Harvard. But during his lectures on scientific inference his colleagues appear to have dozed off. Fisher's 5 percent philosophy was eventually codified by the *Publication Manual of the American Psychological Association,* which dictated the erroneous method worldwide to thousands of academic journals in psychology, education, and related sciences, including forensics.

12. Psychometrics Lacks Power   131

"Power" is a neglected statistical offset to the "first kind of error" of null-hypothesis significance testing. Power assigns a likelihood to the "second kind of error," that of undue gullibility. The leading journals of psychometrics have had their power examined by insiders to the field. The power of most psychological science in the age of Fisher turns out to have been embarrassingly low or, in more than a few cases, spuriously "high"—as was found in a seventy-thousand-observation examination of the matter. Like economists the psychologists developed a fetish for testimation and wandered away from powerful measures of oomph.

13. The Psychology of Psychological Significance Testing 140

Psychologists and economists have said for decades that people are "Bayesian learners" or "Neyman-Pearson signal detectors." We learn by doing and staying alert to the signals. But when psychologists and others propose to test those very hypotheses they use Fisher's Rule of Two. That is, they erase their own learning and power to detect the signal. They seek a foundation in a Popperian falsificationism long known to be philosophically dubious. What in logic is called the "fallacy of the transposed conditional" has grossly misled psychology and other sizeless sciences. An example is the overdiagnosis of schizophrenia.

14. Medicine Seeks a Magic Pill 154

We found that medicine and epidemiology, too, are doing damage with Student's t—more in human terms perhaps than are economics and psychology. The scale along which one would measure oomph is very clear in medicine: life or death. Cardiovascular epidemiology, to take one example, combines with gusto the fallacy of the transposed conditional and the sizeless stare of statistical significance. Your mother, with her weak heart, needs to know the oomph of a treatment. Medical testimators aren't saying.

15. Rothman's Revolt 165

Some medical editors have battled against the 5 percent philosophy. But even the *New England Journal of Medicine* could not lead medical research back to William Sealy Gosset and the promised land of real science. Neither could the International Committee of Medical Journal Editors, though covering worldwide hundreds of journals. Kenneth Rothman, the founder of *Epidemiology*, forced change in his journal. But only his journal. Decades ago a sensible few in education, ecology, and sociology initiated a "significance test controversy." But grantors, journal referees, and tenure committees in the statistical sciences had faith that probability spaces can judge—the "judgment" merely that $p < .05$ is "better" for variable X than $p < .11$ for variable Y. It's not. It depends on the oomph of X and Y.

16. On Drugs, Disability, and Death 176

The upshot is that because of Fisher's standard error you are being given dangerous medicines, and are being denied the best medicines. The Centers for Disease Control is infected with p-values in a grant, for example, to study drug use in Atlanta. Public health has been infected, too. An outbreak of salmonella in South Carolina was studied using significance tests. In consequence a good deal of the outbreak was ignored. In 1995 a Cancer Trialists' Collaborative Group came to a rare consensus on effect size: ten different studies agreed that a certain drug for treating prostate cancer can increase patient survival by 12 percent. An eleventh study published in the *New England Journal of Medicine* dismissed the drug. The dismissal was based not on effect size bounded by confidence intervals based on what Gosset called "real" error but on a single p-value only, indicating, the Fisherian authors believed, "no clinically meaningful improvement" in survival.

17. Edgeworth's Significance 187

The history of this persistent but mistaken practice is a social study of science. In 1885 an eccentric and brilliant Oxford don, Francis Ysidro Edgeworth, coined the very term *significance*. Edgeworth was prolific in science and philosophy, but was especially interested in watching bees and wasps. In measuring their behavioral differences, though, he focused on the sizes and meanings of the differences. He never depended on *statistical* significance.

18. "Take 3σ as Definitely Significant": Pearson's Rule   193

By contrast, Edgeworth's younger colleague in London, the great and powerful Karl Pearson, used "significance" very heavily indeed. As such things were defined in 1900 Pearson was an advanced thinker—for example, he was an imperialist and a racist and one of the founding fathers of neopositivism and eugenics. Seeking to resolve a tension between passion and science, ethics and rationality, Pearson mistook significance for "revelations about the objective world." In 1901 he believed 1.5 to 3 standard deviations were "definitely significant." By 1906, he tried to codify the sizeless stare with a Rule of Three and tried to teach it to Gosset.

19. Who Sits on the Egg of *Cuculus Canorus?* Not Karl Pearson   203

Pearson's journal, *Biometrika* (1901–), was for decades a major nest for the significance mistake. An article on the brooding habits of the cuckoo bird, published in the inaugural volume, shows the sizeless stare at its beginnings.

20. Gosset: The Fable of the Bee   207

Gosset revolutionized statistics in 1908 with two articles published in this same Pearson's journal, "The Probable Error of a Mean" and "The Probable Error of a Correlation Coefficient." Gosset also independently invented Monte Carlo analysis and the economic design of experiments. He conceived in 1926 the ideas if not the words of "power" and "loss," which he gave to Egon Pearson and Jerzy Neyman to complete. Yet most statistical workers know nothing about Gosset. He was exceptionally humble, kindly to other scientists, a good father and husband, altogether a paragon. As suits an amiable worker bee, he planted edible berries, blew a pennywhistle, repaired entire, functioning fishing boats with a penknife, and—though a great scientist—was for thirty-eight years a businessman brewing Guinness. Gosset always wanted to answer the how-much question. Guinness needed to know. Karl Pearson couldn't understand.

21. Fisher: The Fable of the Wasp   214

The tragedy in the fable arose from Gosset the bee losing out to R. A. Fisher the wasp. All agree that Fisher was a genius. Richard Dawkins calls him "the greatest of Darwin's successors." But Fisher was a genius at a certain kind of academic rhetoric and politics as much as at mathematical statistics and genetics. His ascent came at a cost to science—and to Gosset.

22. How the Wasp Stung the Bee and Took over Some Sciences   227

Fisher asked Gosset to calculate Gosset's tables of *t* for him, gratis. He then took Gosset's tables, copyrighted them for himself, and in the journal *Metron* and in his *Statistical Methods for Research Workers,* later to be published in thirteen editions and many languages, he promoted his own circumscribed version of Gosset's test. The new assignment of authorship and the faux machinery for science were spread by disciples and by Fisher himself to America and beyond. For decades Harold Hotelling, an important statistician and economist, enthusiastically carried the Fisherian flag. P. C. Mahalanobis, the great Indian scientist, was spellbound.

23. Eighty Years of Trained Incapacity: How Such a Thing Could Happen   238

R. A. Fisher was a necessary condition for the standard error of regressions. No Fisher, no lasting error. But for null-hypothesis significance testing to persist in the face of its logical and practical difficulties, something else must be operating. Perhaps it is what Thorstein Veblen called "trained incapacity," to which might be added what Robert Merton called the "bureaucratization of knowledge" and what Friedrich Hayek called the "scientistic prejudice." We suggest that the sizeless sciences need to reform their scientistic bureaucracies.

24. What to Do   245

What, then? Get back to size in science, and to "real error" seriously considered. It is more difficult than Fisherian procedures, and cannot be reduced to mechanical procedures. How big is big is a necessary question in any science and has no answer independent of the conversation of scientists. But it has the merit at least of being relevant to science, business, and life. The Fisherian procedures are not.

Preface

The implied reader of our book is a significance tester, the keeper of numerical things. We want to persuade you of one claim: that William Sealy Gosset (1876–1937)—aka "Student" of Student's t-test—was right and that his difficult friend, Ronald A. Fisher, though a genius, was wrong. Fit is not the same thing as importance. Statistical significance is not the same thing as scientific finding. R^2, t-statistic, p-value, F-test, and all the more sophisticated versions of them in time series and the most advanced statistics are misleading at best.

No working scientist today knows much about Gosset, a brewer of Guinness stout and the inventor of a good deal of modern statistics. The scruffy little Gosset, with his tall leather boots and a rucksack on his back, is the heroic underdog in our story. Gosset, we claim, was a great scientist. He took an economic approach to the logic of uncertainty. For over two decades he quietly tried to educate Fisher. But Fisher, our flawed villain, erased from Gosset's inventions the consciously economic element. We want to bring it back.

We lament what could have been in the statistical sciences if only Fisher had cared to understand the full import of Gosset's insights. Or if only Egon Pearson had had the forceful personality of his father, Karl. Or if only Gosset had been a professor and not a businessman and had been positioned therefore to offset the intellectual capital of Fisher.

But we don't consider the great if mistaken Fisher and his intellectual descendants our enemies. We have learned a great deal from Fisher and his followers, and still do, as many have. We hope you, oh significance tester, will read the book optimistically—with a sense of how "real" significance can transform your science. Biometricians who study AIDS and economists who study growth policy in poor countries are causing damage with a broken statistical instrument. But wait: consider the progress we can make if we fix the instrument.

Can so many scientists have been wrong over the eighty years since 1925? Unhappily, yes. The mainstream in science, as any scientist will tell you, is often wrong. Otherwise, come to think of it, science would be complete. Few scientists would make that claim, or would want to. Statistical significance is surely not the only error in modern science, although it has been, as we will show, an exceptionally damaging one. Scientists are often tardy in fixing basic flaws in their sciences despite the presence of better alternatives. Think of the half century it took American geologists to recognize the truth of drifting continents, a theory proposed in 1915 by—of all eminently ignorable people—a German meteorologist. Scientists, after all, are human. What Nietzsche called the "twilight of the idols," the fear of losing a powerful symbol or god or technology, haunts us all.

In statistical fields such as economics, psychology, sociology, and medicine the idol is the test of significance. The alternative, Gossetian way is a uniformly more powerful test, but it has been largely ignored. Unlike the Fisherian idol, Gosset's approach is a rational guide for decision making and easy to understand. But it has been resisted now for eighty years.

Our book also addresses implied readers outside the statistical fields themselves such as intellectual historians and philosophers of science. The history and philosophy of applied statistics took a wrong turn in the 1920s, too. In an admittedly sketchy way—Ziliak himself is working on a book centered on Gosset—we explore the philosophy and tell the history here. We found that the recent historians of statistics, whom we honor in other matters, have not gotten around to Gosset. The historiography of "significance" is still being importantly shaped by R. A. Fisher himself four decades beyond the grave. It is known among sophisticates that Fisher took pains to historicize his prejudices about statistical methods. Yet his history gave little credit to other people and none to those who in the 1930s developed a decision-theoretic alternative to the Fisherian routine. Since the 1940s most statistical theorists, particularly at the advanced level, have not mentioned Gosset. With the notable exception of Donald MacKenzie, a sociologist and historian of science, scholars have seldom examined Gosset's published works. And it appears that no one besides the ever-careful Egon S. Pearson (1895–1980) has looked very far into the Gosset archives—and that was in 1937–39 for the purpose of an obituary.

The evidence on the Gosset-Fisher relationship that Ziliak found in the archives is startling. In brief, Gosset got scooped. Fisher's victory over

Gosset has been so successful and yet so invisible that a 2006 publication on *anti*-Fisherian statistics makes the usual mistake, effectively equating Fisher's approach with Gosset's (Howson and Urbach 2006, 133). In truth it was Gosset, in 1905, not Neyman, in 1938, who gave "the first emphasis of the behavioralistic outlook in statistics" (Savage 1954, 159).

Only slowly did we realize how widespread the standard error had become in sciences other than our home field of economics. Some time passed before we systematically looked into them. Thus the broader intervention here. We couldn't examine every science or subfield. And additional work remains of course to be done, on significance and other problems of testing and estimation. Some readers, for example, have asked us to wade in on the dual problems of specification error and causality. We reply that we agree—these are important issues—but we couldn't do justice to them here.

But we think the methodological overlaps in education and psychology, economics and sociology, agriculture and biology, pharmacology and epidemiology are sufficiently large, and the inheritance in them of Fisherian methods sufficiently deep, that our book can shed some light on all the t-testing sciences. We were alarmed and dismayed to discover, for example, that supreme courts in the United States, state and federal, have begun to decide cases on the basis of Fisher's arbitrary test. The law itself is distorted by Fisher. Time to speak up.

We invite a general and nontechnical reader to the discussion, too. If he starts at the beginning and reads through chapter 3 he will get the main point—that oomph, the difference a treatment makes, dominates precision. The extended but simple "diet pill example" in chapter 3 will equip him with the essential logic and with the replies he'll need to stay in the conversation. Chapter 17 through to the end of the book provides our brief history of the problem and a sketch of a solution.

Readers may find it strange that two historical economists have intruded on the theory, history, philosophy, sociology, and practice of hypothesis testing in the sciences. We are not professional statisticians and are only amateur historians and philosophers of science. Yet economically concerned people have played a role in the logic, philosophy, and dissemination of testing, estimation, and error analysis in all of the sciences from Mill through Friedman to Heckman. Gosset himself, we've noted, was a businessman and the inventor of an economic approach to uncertainty. Keynes wrote *A Treatise on Probability* (1921), an important if somewhat neglected book on the history and foundations of probability theory.

Advanced empirical economics, which we've endured, taught, and written about for years, has become an exercise in hypothesis testing, and is broken. We're saying here that the brokenness extends to many other quantitative sciences—though notably—we could say significantly—not much to physics and chemistry and geology. We don't claim to understand fully the sciences we survey. But we do understand their unhappy statistical rhetoric. It needs to change.

Acknowledgments

We thank above all Morris Altman, who organized a session at the American Economic Association meetings in San Diego on these matters (January 2004) and then edited the articles into a special issue of the *Journal of Socio-Economics* (no. 5, 2004). We have benefited over the years from the comments of a great many scientists, by no means all of them favorable to our views: Theodore W. Anderson, Kenneth Arrow, Orley Ashenfelter, Howard Becker, Yakov Ben-Haim, Mary Ellen Benedict, Nathan Berg, Kevin Brancato, James Buchanan, Robert Chirinko, Ronald Coase, Kelly DeRango, Peter Dorman, Paul Downward, Roderick Duncan, Graham Elliott, Deborah Figart, William Fisher, Edward Fullbrook, Andrea Gabor, Marc Gaudry, Robert Gelfond, Gerd Gigerenzer, Arthur Goldberger, Clive Granger, Daniel Hamermesh, Wade Hands, John Harvey, Reid Hastie, David Hendry, Kevin Hoover, Joel Horowitz, Sanders Korenman, William Lastrapes, Tony Lawson, Frederic Lee, Geoffrey Loftus, Peter Lunt, John Lyons, Andrew Mearman, Peter Monaghan, John Murray, Anthony O'Brien, David F. Parkhurst, John Pencavel, Gregory Robbins, William Rozeboom, David Ruccio, Thomas Schelling, Allan Schmid, George Selgin, Jeffrey Siminoff, John Smutniak, Gary Solon, Dwight Steward, Stephen Stigler, Diana Strassman, Lester Telser, Bruce Thompson, Erik Thorbecke, Geoffrey Tilly, Andrew Trigg, Gordon Tullock, Jeffrey Wooldridge, Allan Würtz, and James P. Ziliak.

Arnold Zellner, the late William Kruskal (1919–2005), Daniel Klein, Stephen Cullenberg, Kenneth Rothman, Edward Leamer, and the late Jack Hirshleifer (1925–2005) have our special thanks. Zellner, Kruskal, Rothman, Leamer, and Hirshleifer have long advocated sanity in significance. It has been inspiriting to have such excellent scientists saying to us, "Yes, after all, you are quite right." We would like especially to thank Arnold

Zellner for plying us with papers and books on Jeffreys's and Bayes's methods that we clearly needed to read. And we thank him, Kenneth Rothman, Regina Buccola, Roger Chase, Charles Collings, Joel Horowitz, Geoffrey Loftus, Shirley Martin, Stephen Meardon, Bruce Thompson, Erik Thorbecke, and the two reviewers, Peter Boettke at George Mason and Julian Reiss, who teaches in Spain, for reading and commenting on substantial parts of the manuscript. Late in the project Pete Boettke saw a need for a wider sociological explanation of an eighty-year-old mistake in science. Thus the penultimate chapter.

Collectively speaking we have been telling versions of our story for some decades now. McCloskey has been dining out on the idea of economic versus statistical significance for over twenty years. J. Richard Zecher first explained the point to her in the early 1980s when they were colleagues at the University of Iowa working on an article on the gold standard. Eric Gustafson had explained it to her when she was an undergraduate at Harvard, but after her "advanced" econometric training in Fisherian methods, vintage 1965, the point slipped away. In 1983 Harry Collins introduced her to the "significance test controversy" in psychology and sociology. She remembers a presentation to a large audience at the American Economic Association meetings in Dallas in 1984, with Edward Leamer and the late Zvi Griliches commenting; and smaller but still crucial seminars at Groningen, Oxford, and the LSE in 1996. The results of all this patient tuition, 1962–96, show up in her *The Rhetoric of Economics* (1985b [1998]).

Ziliak, too, has been the recipient of many courtesies. He first learned of the point in 1988, from the elementary book by Ronald and Thomas Wonnacott (1982), while working in cooperation with the U.S. Department of Labor as a labor market analyst for the Indiana Department of Employment and Training Services. When Ziliak pointed out to the chief of his division that black teenage unemployment rates were being concealed from public view he encountered puzzling resistance. Given the small sample sizes, the chief said, the unemployment rates did not reach an arbitrary level of statistical significance. But the Department of Labor, which authorizes the distribution of official labor market statistics, appeared to be saying that an average 30 or 40 percent rate of unemployment was not discussable because the p-values exceeded .10, the department's shut-up point. Ziliak was embarrassed to return to the telephone to deliver the news to the citizen whose call had started the inquiry. "Sorry, sir. We do not have any quantitative information about black teenage unemployment in the cities." In 1989 he read the first edition of McCloskey's *The Rhetoric of Economics* (1985b [1998]), including the then startling chapters on "significance." Two years later he moved to Iowa and the graduate study of economics, where soon McCloskey invited him to join forces. Talks given jointly with McCloskey at Iowa (1993, 1994) and then solo at Indiana (1995) and the Eastern Economic Association (New York, 1995) transformed early puzzlement into action.

Throughout the 1980s and 1990s, then, we were talking and talking, individually and as a tag team, persuading a happy few. In recent years (we sense we have not mentioned all the events and apologize) we have found often appreciative and always attentive and sometimes stunned audiences at the annual meetings of the American Economic Association (Chicago, 1998; San Diego, 2004), Ball State University, Baruch College (School of Public Affairs), Bowling Green State University (a student seminar), the bi-ennial meetings of the Association for Heterodox Economics (University of Leeds, 2004), the Association for Heterodox Economics Post-graduate Workshop on Research Methods (University of Manchester, 2005), the University of Chicago (Center for Population Economics, 2005), the University of Colorado-Boulder, Dennison University, the Eastern Economic Association/Association for Social Economics (New York, 2003), the Elgin Community College/Roosevelt University Faculty Speaker Series, Erasmus University of Rotterdam, the summer institutes over many years of the European Doctoral Association in Management and Business, the First International Congress of Heterodox Economics (University of Missouri, Kansas City, 2003), George Mason University (Philosophy, Politics, and Economics Seminar), the University of Georgia, the Georgia Institute of Technology, Göteborg University (twice), Harvard University (a seminar for graduate students in economics; the faculty was skeptical), the University of Illinois at Chicago, Illinois State University, Macquarie University (Australia), the University of Michigan (another student seminar), the University of Nebraska, Northwestern University (Economic History Workshop), the University of Wisconsin (still another student seminar; the faculty was outraged), the Rhetoric and Economics Conference (organized by Paul Turpin at Milliken University, 2005), and annual meetings of the Southern Economic Association (New Orleans, 2004), the Ratio Institute of Stockholm (2006), and the University of Wollongong.

Ziliak gratefully acknowledges the cooperation of libraries and their staffs: University College London, Special Collections, where Gillian Furlong and Steven Wright gave able and kind access to the Galton Papers and Pearson Papers (containing files on Karl Pearson, Egon Pearson, and Gosset, Fisher, and Neyman); the Guinness Archives, Diageo (Guinness Storehouse, Dublin, where Eibhlin Roche and Clare Hackett are themselves a "storehouse" of ideas); the Museum of English Rural Life, University of Reading; the University of Illinois at Chicago, Special Collections, Richard J. Daley Library, Science Library, Mathematics Library, and Health Sciences Library; the University of Chicago's Eckhart Library (for providing access to *Letters of William Sealy Gosset to R. A. Fisher, 1915–1936, Vols. 1–5* (private circulation, 1962) and its Regenstein, Yerkes, and Crerar libraries; Roosevelt University's Murray-Green Library; Emory University's Woodruff Library; and the libraries at the Georgia Institute of Technology, the University of Iowa, and Bowling Green State University. For research assistance at various stages of the project we were fortunate to employ Cory Bilton, Angelina Lott, David McClough, and Noel Winter.

Ziliak also thanks the Institute for Humane Studies for a Hayek Scholar Travel Grant and Roosevelt University for two summer grants used to collect primary materials in London, Dublin, Reading, and Chicago. Roosevelt is a rare site of sanity in academic life, serious about justice and freedom. He thanks there many colleagues who have tolerated his brief lectures on significance over lunch, especially Stefan Hersh, a friend who combines oomph *and* precision and Lynn Weiner and Paul Green, for openhandedly helping. In London Ziliak was well cared for, too. Andrew Trigg and his sons provided an amusing diversion from Gower Street, and Sheila Trigg, an Oxford-trained political adviser and chef extraordinaire, was her usual goddess self. McCloskey thanks the College of Liberal Arts and Sciences at the University of Illinois at Chicago for continuing research moneys used for the project and her colleagues at UIC, such as Lawrence Officer, who have Gotten It.

We also gratefully acknowledge the University of Wisconsin Press, the *Journal of Economic Literature,* the *Journal of Socio-Economics,* and *Rethinking Marxism* for permission to use some of our earlier writings and statistics; University College London, Special Collections Library, for permission to quote from the Galton Papers and Pearson Papers; Guinness Archives (Diageo) for permission to reproduce images of Gosset; the *Journal of Socio-Economics* and Professor Erik Thorbecke for permission to print a version of Thorbecke's (2004) very illuminating figure on economic significance; *Biometrika,* for permission to reprint a page from Student 1908a; *Educational and Psychological Measurement* for permission to reprint a table from Fidler et al. (2004b); the Johns Hopkins University School of Hygiene and Public Health and the *American Journal of Epidemiology* for permission to reproduce a table from Savitz, Tolo, and Poole (1994); the *New England Journal of Medicine* for allowing a version of a figure from Freiman et al. (1978); and Professor Kenneth Rothman for supplying an unpublished graph of a p-value function. James F. Reische, our editor at the University of Michigan Press, carried our book over the hurdles.

Ziliak would be lost without Flora, Jude, and Suzette, whose love is all about oomph. And he dedicates the book to his parents, Barbara and Lawrence Ziliak, real world examples of unconditional love. McCloskey dedicates the book to her grandchildren, Connor and Lily. May they someday read this and understand a part of love.

Love comes in more academic forms, too: together we dedicate the book to the memory of William H. Kruskal for his many kindnesses extended from the 1970s to the 2000s and for a long life in theoretical and applied statistics of substantive significance.

The Cult of Statistical Significance

A Significant Problem

Merely theoretical uncertainty continues to have no meaning. . . .
Perhaps as near to it as we can come is in the familiar story of the
Oriental potentate who declined to attend a horse race on the
ground that it was already well known to him that one horse could
run faster than another. His uncertainty as to which of several
horses could outspeed the others may be said to have been purely in-
tellectual. But also in the story nothing depended from it; no cu-
riosity was aroused. . . . In other words, he did not care; it made no
difference. And it is a strict truism that no one would care about
any exclusively theoretical uncertainty or certainty. For by defini-
tion in being exclusively *theoretical it is one which makes no dif-*
ference anywhere.

JOHN DEWEY 1929, 38–39. EMPHASIS IN ORIGINAL

For the past eighty years it appears that some of the sciences have made
a mistake by basing decisions on statistical "significance." Although it
looks at first like a matter of minor statistical detail, it is not.

Statistics, magnitudes, coefficients are essential scientific tools. No one
can credibly doubt that. And mathematical statistics is a glorious social
and practical and aesthetic achievement. No one can credibly doubt that
either. Human understanding of chance and uncertainty would be much
reduced were it not for Bayes's rule, gamma functions, the bell curve, and
the rest. From the study of ancient parlor games to the rise of modern
space science, mathematical statistics has shown its power. Our book is
not a tract against counting or statistics. On the contrary. In our own scientific work we are quantitative economists and value statistics as a crucial tool.

But one part of mathematical statistics has gone terribly wrong, though mostly unnoticed. The part we are worrying about here seems to have all the quantitative solidity and mathematical shine of the rest. But it also seems—unless we and some other observers of mathematical statistics such as Edgeworth, Gosset, Egon Pearson, Jeffreys, Borel, Neyman, Wald, Wolfowitz, Yule, Deming, Yates, Savage, de Finetti, Good, Lindley, Feynman, Lehmann, DeGroot, Bernardo, Chernoff, Raiffa, Arrow, Blackwell, Friedman, Mosteller, Tukey, Kruskal, Mandelbrot, Wallis, Roberts, Granger, Press, Moore, Berger, Freedman, Rothman, Leamer, and Zellner are quite mistaken—that reducing the scientific problems of testing and measurement and interpretation to one of "statistical significance," as some sciences have done for more than eighty years, has been an exceptionally bad idea.

Statistical significance is, we argue, a diversion from the proper objects of scientific study. Significance, reduced to its narrow statistical meaning only, has little to do with a defensible notion of scientific inference, error analysis, or rational decision making. And yet in daily use it produces unchecked a large net loss for science and society. Its arbitrary, mechanical illogic, though currently sanctioned by science and its bureaucracies of reproduction, is causing a loss of jobs, justice, profit, and even life.

We and our small (if distinguished) group of fellow skeptics say that a finding of "statistical" significance, or the lack of it, statistical *in*significance, is on its own almost valueless, a meaningless parlor game. Statistical significance should be a tiny part of an inquiry concerned with the size and importance of relationships. Unhappily it has become the central and standard error of many sciences.

SIGNIFICANCE IN SCIENCE

Statistical significance is the main factual tool of medicine, economics, psychiatry, agronomy, pharmacology, sociology, education, some parts of the biological and earth sciences, and some parts of the academic study of business and history. Astronomers use it to shine a light. Psychologists have developed a fetish for its scientific-sounding rituals. Poll takers and market analysts depend on little else. The tool and its rituals are not much used in the other sciences—atomic physics, say, or cell biology or chemistry or the remaining parts of the life, earth, atmospheric, or historical sciences.

Statistical significance developed first—if comparatively informally—in demography and astronomy. Toward the end of the eighteenth century Laplace began to formalize the notion of significance in the astronomical sciences, though we know from Elizabeth Scott (1953) and others that it functioned in the astronomy of his day as a tiny supplement to scientific reasoning. By the end of the nineteenth century "significance" had been further mathematically refined and extended and began to shed its uncertain light on many fields, both experimental and observational. It became for instance through the works of Galton, Pearson, and Weldon the main instrument of large-sample biometrics.

A brewer of beer, William Sealy Gosset (1876–1937), proved its value in *small* sample situations. He worked at the Guinness Brewery in Dublin, where for most of his working life he was the head experimental brewer. He saw in 1905 the need for a small-sample test because he was testing varieties of hops and barley in field samples with N as small as four. Gosset, who is hardly remembered nowadays, quietly invented many of the tools of modern applied statistics, including Monte Carlo analysis, the balanced design of experiments, and, especially, Student's t, which is the foundation of small-sample theory and the most commonly used test of statistical significance in the sciences.[1] Gosset's "The Probable Error of a Mean" was published in 1908 under the pseudonym "Student." Yet he had been thinking about his need for a small sample test since at least 1905, the year he told Karl Pearson about it. But the value Gosset intended with his test, he said without deviation from 1905 until his death in 1937, was its ability to sharpen statements of *substantive* or *economic* significance. Gosset's immediate goal, of course, was to brew the best tasting stout at a satisfying price. His test of statistical significance could, he knew, contribute only a little to those substantive aesthetic and economic goals. Experiments in the selection and cultivation of barley and hops, and in quantitative simulations, technologies of malting, water quality, yeast chemistry, storage temperature, cask type, and many other Guinness variables—from an unusually generous if paternalistic wages and benefits scheme to the daily taste test—would contribute far more.[2] World War I had been under way for more than a year when Gosset—who wanted to serve in the war but was rejected because of nearsightedness—wrote to his elderly friend, the great Karl Pearson: "My own war work is obviously to brew Guinness stout in such a way as to waste as little labor and material as possible, and I am hoping to help to do something fairly creditable in that way."[3] It seems he did.

COUNTING MATTERS, AND INFERENCE, TOO

Every science uses counting and should. Counting is central to a real science, and applied statistics is sophisticated counting. The big scientific question is, "How much?" To answer the how-much question you will often need statistical methods. If your clinical-diagnostic problem can best be answered with "analysis of variance"—though we advise you to think twice—then you had better not be terrified of its imposing columns and nonlinear off-diagonals. If your business problem is a sampling one, and requires a high degree of confidence in the truth of the result, you had better examine the entire "power function." If your physical problem leads naturally to the Poisson distribution you had better be fluent in that bit of statistical theory. If you want to say how much abortion reduces crime rates, "other things equal," you had better know how to run multiple regressions that can isolate the effect of abortion from those other things. Good.

Statistical significance is a subset of such statistical methods. Formally speaking, statistical significance is a subset of induction, either "inductive behavior" (the question of how much) or "inductive inference" (the question of whether), depending on one's philosophical school of thought.[4] But statistical "significance," once a tiny part of statistics, has metastasized. You can spot statistical significance in the sciences that use it by noting the presence of an F or p or t or R^2—an asterisk superscripted on a result or a parenthetical number placed underneath it, usually with the word *significance* or *standard error* in attendance. Scientists use statistical significance to "test" a hypothesis—for example, the hypothesis that comets come from outside the solar system or the hypothesis that social welfare programs diminish the pace of economic growth.[5] So statistical significance is also a subset of testing.

Testing, too, is used by all the sciences, and of course should be. That is, claims should be estimated and tested. Scientific assertions should be confronted quantitatively with the world as it is or else the assertion is a philosophical or mathematical one, meritorious no doubt in its own terms but not scientific. To demonstrate scientifically and statistically that Chicago is the City of the Big Shoulders you would need to show by how much Boston or London lack such shoulders.

The problem we are highlighting is that the so-called test of statistical significance does not in fact answer a quantitative, scientific question. Statistical significance is not a *scientific* test. It is a philosophical, qualitative test. It does not ask how much. It asks "whether." Existence, the question of whether, is interesting. But it is not scientific.

The question of whether is studied systematically in departments of philosophy or mathematics or theology. We have spent time in those departments and have a high opinion of them. But their enterprises are not scientific. Instead of asking scientifically how big *are* the shoulders of Chicago, the philosophical disciplines ask *whether* big shoulders *exist*. Does there exist an obligation to believe a synthetic proposition? Yes or no. Does there exist a good and omnipotent God? Yes or no. Does there exist an even number not the sum of two primes? Yes or no. Does there exist a significant relationship between economic growth and belief in hell? Pray tell.

The problem we are pointing out here—we note again that it is well known by sophisticated students of the matter and is extremely elementary—is that by using Fisherian methods some of the putatively quantitative sciences have slipped into asking qualitatively whether there *exists* an effect of drug prices on addiction or whether there *exists* an effect of Vioxx on heart attacks or whether there *exists* an effect of Catholicism on national economic backwardness. Yes or no, they say, and then they stop. They have ceased asking the scientific question "How much is the effect?" And they have therefore ceased being interested in the pragmatic questions that follow: "What Difference Does the Effect Make?" and "Who Cares?" They have become, as we put it, "sizeless."

But the Point of Counting and Inference Is to Find a Size

Real science depends on size, on magnitude. Scientific departments of physics, economics, engineering, history, medicine, and so forth intend to study actualities and realistic possibilities quantitatively. We admire their quantitative intentions, in many fields first imagined in part by statistical sophisticates such as Francis Galton and Karl Pearson. Victor Hilts, the historian of science, wrote in 1973, "I think it might even be fair to say that the introduction of statistical techniques in the social sciences represents one of the two most important methodological innovations in [all of] nineteenth-century science"(Hilts 1973, 207). We agree.

But after Galton and Pearson, and especially after Ronald Fisher, the statistical sciences have slipped into asking a philosophical and qualitative question about existence instead. The scientific question is how much this particular bridge, or a bridge of this particular kind, can tolerate thus-and-such forces of stress. There may "exist" a stable bridge. But unless the magnitudes and limits of stability can be given quantitatively in the world we actually inhabit the knowledge of whether it exists is unhelpful. No astronomer is interested in the question of whether there is some effect of the rest of the galaxy's gravitation on the Oort cloud. No scientific brewer of Guinness will ask whether bitterness "exists"—as a careful student of hops chemistry, and a profit center, he is forced to ask how much. The question of whether has, as John Dewey observed, "no [scientific] meaning," no big bang. Being "exclusively theoretical" no curiosity is aroused by it because it makes "no difference" anywhere. Not even in philosophy, pragmatically considered, the great philosopher said.

Like an engineer the astronomer is interested in *how much* the effect on the Oort cloud is, for the generation of comets, say. Instead of *nonexistent* or *not statistically significant* she uses a word seldom heard in the ontological and metaphysical departments of philosophy and mathematics and theology: *negligible*. In a department of mathematics it would be viewed as irrelevant, even vulgar, to note that the number of even numbers not expressible as the sum of two primes is entirely negligible—as in fact it appears to be. No such even number has yet been found, up to gigantic numbers. But that, the mathematician will complain with a sneer, is a mere calculation, satisfactory for mere engineers and physicists but not for a "real" mathematician. The real mathematician, since the Greeks first invented such a character, has craved certitude. But certitude is not interesting to the astronomer. She seeks magnitude and effect size. She is seldom tempted to substitute "testing for low magnitude" or "estimation of effect size" for "significance testing for a low probability." In a science, size matters.

To confront her assertions with the world the astronomer or engineer uses simulation methods, for which often enough no unique analytic solution is guaranteed to "exist." She is not against analytic certitude, whatever that might mean. Her quantitative methods—estimating magnitudes—puzzle her colleagues in mathematics and, more to the point here, puzzle, too, her colleagues in econometrics or statistical medicine who have spent too much time in the exist/not-exist world of esteemed departments of mathematics, philosophy, or theology. The astronomical scientist wants to meet scientific, not metaphysical, standards. She seeks salience, adequacy, nonnegligibility, real error, an oomph that measures the practical difference something makes. Such scientific standards of persuasion have little to do with axiom or consistency or existence. Axiom, consistency, and existence are the values of the nonscientific departments, admirable in themselves, we repeat earnestly. But they are not the values of a quantitative science.

The substitution of existence for magnitude has been a grave mistake. That's one way of stating the main point of our book. Any quantitative science answers how much, or should. Many do, like geophysics and chemistry and a good deal of history. How much energy did the crashing of the Indian subcontinent contribute to the raising of the Himalayas? How much did foreign trade contribute to the British industrial revolution? How much genetic material is transmitted to the next generation? But medicine, economics, and some other sciences have stopped asking how much, especially in their academic, as against their applied, work. Or, to be more exact, they believe they *are* interested in the quantitative questions of what this or that number really is in the world. But their way of deciding what that number really is—statistical significance or insignificance—and what difference it makes—doesn't give them the correct answer.

Statistical significance sounds scientific. After all, it speaks in technical terms about the experimental and observational quantities of science. And it shows up in, for example, the technical notes to a bottle of prescription pills. "Zyprexa was significantly better than placebo ($P = 0.05$) on initial assessment," says the guardian of the quantitative bottom line behind an antidepressant pill, "with the Cox model but not when compared with placebo by means of a Hochberg adjustment for multiple comparisons" (Osterweil 2006, 2). How very significant.

But in truth statistical significance is a philosophy of mere existence. And even by philosophical standards—leaving aside the scientific standard—it is a poor one. It concerns itself only with one kind of probability of a (allegedly) randomly sampled event—the so-called exact p-value or Student's t—and not with the other kinds of sampling probability, such as the "power of the test," which controls for what Egon Pearson called in 1928 the "second type of error." And statistical significance is not concerned with any of the long list of *nonsampling* sources of error, such as confounding effects, as one finds in medicine and epidemiology; specification error, as one finds in economics and the other human sciences; "non-linear fertility slopes" (Student 1938, 374), as one finds in agronomic experiments; the "bias of the auspices" (Deming 1950, 43), as one finds in government, industrial, and ethnographic studies; measurement error, as one finds in psychology and pharmacology research and everywhere else; or experimental error and sample selection bias, as again one finds in all of the sciences.[6] Significance is a strikingly partial philosophical account of the existence of error, and Fisher's version of it—sans power—is lacking even a probabilistic means of assessing its own magnitude. It fails, as the great geo- and astrophysicist Harold Jeffreys reminded Fisher, to "give us ground for believing the laws that we do believe or else say definitely that our inferences are fallacious."[7]

But the biggest problem is the more elementary one we shall explore here. It is: *Fit is not the same thing as importance.* So-called tests of fit, such as Student's *t* or tests of R^2, do not by themselves solve the all important matter of *Gosset* significance—which is Size Matters + Who Cares?—the "minimax strategy" or "loss function" of every inquiry.[8] Scientists and their customers wish to have relevance not amateur philosophy. They wish for standards that will help them, say, minimize the maximum loss of jobs, income, profit, health, or freedom in following this or that hypothesis as if true.

The general validity of the loss function way of thinking is, incidentally, not sensitive to the degree of risk aversion felt by an individual investigator or his advisees. So "personal taste" or "equal distribution of ignorance" are not excuses justifying gross indulgence of the mechanical instrument. Loss is in the behavioralistic tradition of statistics the perceived value of a sacrifice, giving up that to do this, and may be positive or negative. So even the gambler who sees "nothing but blue sky"—a real risk lover—will distinguish maximum losses from minimum, big wins from small. Adjusting the levels of Type I and Type II error is, statisticians agree, necessary for handling differential attitudes toward risk. The problem is that today's statistical experts do not estimate or consider the loss function or Type II error at all. In his last year of life, the great statistician and economist Leonard Savage asked, "When is one [statistical] expert, real or synthetic, to be preferred to another?" He replied, "Employ, until you have further experience, that expert whose past opinions, applied to your affairs, would have yielded you the highest average income" (1971b, 145–46). Substitute "highest average income"—or rather add to it—other concerns, such as "highest average quality" or "highest rate of patient survival" or "lowest number of heart attacks" or "highest average rate of minority student graduations," or even "highest scientific consensus" and you have what we are claiming here.

In any case, without a loss function a test of statistical significance is meaningless, no better than a table of random numbers. Pretending to afford a view from everywhere, statistical significance is in fact a view from nowhere. In its desire to maximize precision in one kind of sampling error it turns away from the human purposes and problems that motivated the research in the first place. Fisher-significance is by itself about precisely nothing.

Statistical significance has translated every quantitative question about hypotheses into a philosophical and qualitative measure of probability about the data assuming the truth of a singular hypothesis. It has collapsed the scientific world into a Borel space, p (0, 1.0)—a procedure, by the way, that the mathematical statistician Émile Borel (1871–1956) himself emphatically rejected. Borel, though a master of abstract imagination, was deeply interested in the substantive side of testing and in Paris in the 1920s helped convert a young Jerzy Neyman to a life of substantive significance.[9]

Savage noted in *The Foundations of Statistics* (1954) a part of the problem we are highlighting: "Many [scientists following in the footsteps of Karl Pearson and R. A. Fisher]," he wrote, "have thought it natural to extend logic by setting up criteria for the extent to which one proposition tends to imply, or provide evidence for, another. . . . It seems to me obvious, however, that what is ultimately wanted is criteria for deciding among possible courses of action."[10] Yet, in imitation of Fisher, today's significance testers do not think about "possible courses of action." Statistical significance is, as Savage says, "at best a roundabout method of attack."[11] That is to put it charitably.

To cease measuring oomph and its relevant sampling and nonsampling error is to wander off into probability spaces, forgetting—commonly forever—that your interest began in a space of economic or medical or psychological or pharmacological significance. In applied work at, for example, the Pfizer Corporation or the Centers for Disease Control or the Federal Reserve Board of Governors, one would expect the scientists involved to be serious about substantive oomph in medical or economic or pharmacological matters. But p and t and F and other measures of "significance" fill the air.

The sociological pressure to assent to the ritual is great. In 2002 we gave together a talk at the Georgia Institute of Technology, where Ziliak was teaching, on the significance mistake in economics. Three researchers from the nearby Centers for Disease Control (CDC) attended. They agreed with us about "the cult of p," as they put it. But they feared that their mere presence at a lecture against Fisher's "significance" would put their jobs at risk and made us promise not to reveal their names. The official rhetoric at the CDC is: "Second-hand smoke is killing thousands annually. But is it *statistically* significant?" The CDC, which is unquestionably one of the world's most important sites for the scientific study and control of disease, imposes Fisherian orthodoxy on its scientists. That is the power that the cult of statistical significance has.

Sizelessness is not what most of the Fisherians believe they are getting. The sizeless scientists have adopted a method of deciding which numbers are significant that has little to do with humanly significant numbers. The scientists are counting, to be sure: "3.14159***," they proudly report, or simply "***." But, as the probabilist Bruno de Finetti said, the sizeless scientists are acting as though "addition requires different operations if concerned with pure numbers or amounts of money" (De Finetti 1971, 486, quoted in Savage 1971a).

Substituting "significance" for scientific how much would imply that the value of a lottery ticket is *the chance itself,* the chance 1 in 38,000, say, or 1 in 100,000,000. It supposes that the only source of value in the lottery is sampling variability. It sets aside as irrelevant—simply ignores—the value of the expected prize, the millions that success in the lottery could in fact yield. Setting aside both old and new criticisms of expected utility theory, a prize of $3.56 is very different, other things equal, from a prize of $356,000,000.[12] No matter. Statistical significance, startlingly, ignores the difference.

Imagine that you and your infant child are standing on a sidewalk near a busy street. You have just purchased a hot dog from a street vendor, and have already safely crossed the street. You suddenly realize that you've forgotten the mustard. Prize Number One: if you and your child scurry across the busy street, dodging moving trucks and cars, there is some probability—say, 0.95—you'll both safely return with the mustard in hand.

Now imagine that you've gotten the hot dog and mustard across the street all right—but this time you've forgotten your infant child. You watch in horror as she tries to cross the street by herself. Prize Number Two: if you scurry across the street, dodging vehicles as before, there is some probability—say, the same 0.95—you'll return with your child unharmed.

Two prizes—the mustard and your child—identical probability. Statistical significance ignores the difference. It ignores, in the words of Savage, "criteria for deciding among possible courses of action." Since both decisions are equal in probability of "success," that is, equal in *statistical* significance, the sizeless scientists assign equal value to the mustard and the child. Both the CHILD variable and the MUSTARD variable are significant at $p = .05$. Therefore, the sizeless scientist in effect declares, "They are equally important reasons for crossing the street."

Imagine a crime suspect interrogated repeatedly by investigators. Significance testers say that if ninety-five times or more out of a hundred a suspect admits he is guilty of the crime, the correct decision is "guilty," hang him high, regardless of the size or nature of his alleged crime, ax murder or double parking, and the ethical and pecuniary cost to society of disposing of him. But if his testimony is the same only ninety-four times of out a hundred, or eighty-one times out of a hundred, let the suspect go unmolested. He's innocent, again despite the size, nature, or cost of the alleged crime. If the p doesn't fit—the sizeless scientists say, following the dictates of R. A. Fisher—then you must acquit.

William James believed that "we have the right to believe any hypothesis that is live enough to tempt our will" (1896, 107) but added—in a philosophical pragmatism friendly to what Neyman and Savage called "inductive behavior"—what really tempts is what a belief "leads to" (98). What, a scientist should ask, are the social or personal human purposes activated by the belief? What does the belief lead to? A conclusion such as "the subjects of my inquiry are indifferent between a mustard packet and a child" will rarely lead one to the correct side of the street. A sizeless scientific finding, measuring nothing relevant to human decision making, leads nowhere—it is therefore, in strictest pragmatist terms, as the philosophers Quine and Ullian put it, "unbelievable."[13]

The usual procedure does not ask the question "How big is big?" about its numbers . It does not ask whether the variable is for human or other purposes substantively significant. It asks instead, "In the data we happen to have, is the estimated variable (or the full model, if that's what is being tested) more than two or three standard deviations of its own sampling variation away from the null hypothesis?" It's not the main question of science. But it is the only question that a sizeless scientist bothers with. "X has at the .05 level a significant effect on Y," he says. "Therefore X is important for explaining Y." The circumlocution is made regardless of how much a unit increase in X affects the levels or qualities of Y or how many other variables are involved or how much X varies or what difference X makes in the course of Y when Z is added to the experiment. Dewey wrote that "it is a strict truism that [no scientist] would care about *any* exclusively theoretical uncertainty or certainty. For by definition in being *exclusively* theoretical it is one which makes no difference any-where" (1929, 38–39, emphasis in original). Yes.

When the Fisherian significance tester wishes to use a set of data to distinguish between two different hypotheses, such as $\beta = 1$ (the null hypothesis, say) and $\beta > 1$ (the one-sided alternative), he asks *whether* the variate is "statistically significantly different from the null" (which null hypothesis he assumes provisionally to be true). Exclusively he asks about the theoretical certainty or lack of it. He does not ask the relevant scientific question—how much oomph, how much pragmatic effect, what difference does the variate or model lead to relative to some different magnitude of scientific importance? Does a β of 1.20 lead to other economic or medical recommendations? Does a β of 1.10 tempt our will? What are the available courses of action and how do you know?

In a regression context, an estimated $\beta_{hat} = .01$ (to take another magnitude) is said by such a scientist to be *different from* a theoretical null hypothesis β_{null}, equal to, let's say, exactly zero, if with a large enough sample and a small enough variation in the sample at hand the variate attached to β_{hat} is "statistically significantly different from the null hypothesis of [exactly] zero." That the independent variate in question is "amount of insulin dose" and the dependent variate "length of patient survival" does not affect procedures. The Fisherian procedure claims to test the significance of numbers "in their own terms," objectively, without regard for human purposes. (*In*significance, we should add, a failure to achieve the .05 or .01 cutoff, is the other side. "The coefficient on DIALYSIS [or MONEY or TAXES or DEATH—the list of candidates is endless] is statistically *in*significant but of the right sign," as a great many authors have written in slavish imitation of each other. In our home field of economics we call such practice "sign econometrics." It is rampant.) But the equation and substitution of statistical significance and scientific relevance are proposed relative to no scientific or ethical values. Literally, we repeat, none.

A book could be excellently copyedited, precise in every detail, the publisher and the author having spent months and months, thousands of man hours, making sure that every number in the book is precise to eight significant digits, every jot and tittle of every word just so, a very Torah scroll of precision. Yet the book could be substantively worthless, a book, say, consisting of Fisherian significance tests on numbers gathered with no scientific question in mind from telephone directories and MySpace sites in Holland, Greece, and Tanzania. On the other hand, an important book, such as Fisher's *Design of Experiments* (1935), could be imprecise in very many details, with notable misspellings, say, and errors in the statistical tables, illogicalities in the mathematics, or, worse (as we claim was in fact the case for *The Design of Experiments*), it could have vitiating errors in its scientific rhetoric, yet it could nonetheless be important in the intellectual history of the twentieth century and well worth publishing. No sensible person—or publisher—would confuse precision in production with importance in intellectual life. Yet that is what users and buyers of tests of significance do.

Precision in the matter of random sampling is nice, to be sure, something to be desired. Sometimes having "precision" narrowed to "precision in the sense of an arbitrarily low standard error from an alleged sample of an imaginary repeated experiment in view of an arbitrary null hypothesis of exactly zero" is scientifically useful. But rarely. The problem is the opportunity cost of specializing in it excessively, as Fisherian procedures urge one to do. A publisher would never say to herself, "This book is an idiotic compilation of significance tests on telephone directories and MySpace sites that leads nowhere and tempts no will. But after all it is beautifully and precisely copyedited, with every number and spelling and citation checked fifty times. Therefore I have done my job." But that, alas, is what the sizeless sciences say to themselves, especially since the arrival of the desktop computer with its ability to invert big matrices at the punch of a key, "checking" on sampling variability effortlessly and on a gigantic scale. By Moore's law electronic computation of statistical significance has cheapened to near zero. By economic law the scientific value of the computation has come to be equal at the margin to its private cost. "Decision" has become socialized and bureaucratized—heedless of the social margins. The sciences of medicine and economics and the others have developed machinery heedless of scientific substance.

The substitute question is supposed to tell "whether" an effect "exists." It does not. If beneath your β's you can insert $p < .05$ or $t > 2.0$, then the scientific job is supposed to be finished regardless of effect size and its relevance. It is not.

The Sizeless Scientists Have Missed the Point

The problem of significance is old. The substitution of significance for relevance is more than a century old in some sciences, such as parts of biology—anthropometry, for example. And since Ronald Aylmer Fisher made it canonical in the 1920s and 1930s the standard error has spread and spread. The idea of significance as applied to matters of random sampling existed, we have noted, in the early eighteenth century, originating, it would appear, with John Arbuthnot and Daniel Bernoulli. The very word *significance* seems to have first appeared in an article by F. Y. Edgeworth, in 1885. The basic logic is ancient and has, we've said, its uses. Cicero's Quintus character in *De Divinatione* uses it to argue for the existence of the gods: "'Mere accidents,' you say. . . . Can anything be an 'accident' which bears on it every mark of truth? Four dice are cast and a Venus throw [four different numbers] results—that is chance; but do you think it would be chance, too, if in one hundred casts you made one hundred Venus throws?" (I, 23). He's got a point.

But significance was a minor part of any science until Francis Galton (1822–1911) and especially Karl Pearson (1857–1936) fitted *Biometrika* with chi-square, regressions, and other curves. After Fisher published *Statistical Methods for Research Workers* in 1925, among the most widely read professional texts on statistics of the twentieth century, statistical significance became the central empirical question, commonly the only empirical question—first in biometry and agronomy, then in genetics, psychology, economics, anthropometry, sociology, medicine, law, and many other fields. In psychometrics, statistical significance was by the late 1920s—regardless of effect size—judged a necessary and sufficient condition for the demonstration of a scientific result. By the 1970s Fisher's absolute criterion of significance was flourishing in all the fields we highlight here and was sinking its roots into a few more that we have only glanced at, such as law.

Fisher's procedure of statistical significance often has other difficulties, on which a great deal of philosophical and mathematical and sociological ingenuity has been spent. Is the sample proper? Is the whole population in the so-called sample? Is the "sample" one of convenience, and therefore biased in preselection? Has the investigator herself selected for use only the significant results ($p < .05$) out of many experiments attempted? Is the significance level corrected for sample size? Is the assumed sampling distribution the correct one? What, after all, is the correct measure of probability? Is probability about *belief*—in De Morgan's terms, a "law of thought"—or is it mainly about frequencies in the long run—a "law of things," as John Venn believed?[14] What is the relation between "personal" probability and groupwise calculable risk?[15] What about heteroskedacity? What about truncation error? What about specification error? And so on. You may consult many thousands of excellent philosophical and statistical articles published each year that examine these absorbing difficulties.

But we are making a more elementary point. Statistical significance, we are saying, is never the end of an argument. Indeed, speaking of stepwise procedures, it's always a false start. Statistical significance is neither necessary nor sufficient for a scientific result. Many who have seriously examined the issue, from Edgeworth to Kruskal, agree. Statistical significance offers merely a certain kind of theological proof: ipse dixit—a *t*-statistic is supposed to "speak for itself" because the probability of a certain restricted kind of error is low, a sampling error of excessive skepticism. Most scientists seem to believe this. In grant applications as much as journal correspondence, they verbally repeat the belief as if it were the very paternoster of science. We want to persuade them to go back to a truly scientific ritual, asking How Much.

SCIENCE NEEDS A LOSS FUNCTION

The doctor who cannot distinguish statistical significance from substantive significance, an *F*-statistic from a heart attack, is like an economist who ignores opportunity cost—what statistical theorists call the loss function. The doctors of "significance" in medicine and economy are merely "deciding what to say rather than what to do" (Savage 1954, 159). In the 1950s Ronald Fisher published an article and a book that intended to rid *decision* from the vocabulary of working statisticians (1955, 1956). He was annoyed by the rising authority in highbrow circles of those he called "the Neymanites."

But every inference drawn from a test of statistical significance is a "decision" involving substantive loss and, further, not merely one narrow sort of loss under conditions of random sampling. Every decision involves cost and benefit, needs and wants, choices and courses, a minimax problem (if that is your loss function) as general or particular as the problem-situation warrants. Accepting or rejecting a test of significance without considering the potential losses from the available courses of action is buying a pig in a poke. It is not ethically or economically defensible.

We want you to be dissatisfied with a 5 percent "verbalistic" philosophy (Savage 1954, 159). The solution is not to seek a 1 percent or a 12 percent philosophy. No predetermined rule of one or three or whatever for α-level sampling error will do, as Gosset said repeatedly to Fisher, Karl

Pearson, Egon Pearson, Edwin S. Beaven, and others. Significance rules in isolation are useless.

All the signals point you toward seeking instead, speaking in regression terms, a 100 percent β-philosophy. To reclaim the quantitative side of your science you will need to make β-decisions: you will have to make β-decisions about differences expressed in terms of effect size and, if you are seriously worried about sampling as one source of error, you will have to make β-decisions about your power to reject the null relative to substantive alternatives. You will have to think about your coefficients in a currency of How Much in the world as it is, or could be, and persuade a community of scientists. Instead of deploying a mechanical rule about one kind of sampling error you will have to establish a reservation price of β-coefficients, a minimum effect size of substantive significance, in the relevant range of power, for your particular area of research, acknowledging all the sources of error. You will not confuse power with substance, nor mere sampling error with "real" or "actual" error (Student 1927, 1938). You will instead dwell on the substantive meaning of your estimates in the range of real error. That is, you will *actually* repeat experiments, as Gosset did, not pretend to, as Fisher and his many followers have, so that your sampling distribution is based on something besides an imagined infinitely repeated flipping of a fair coin. You will employ minimax or some other loss function to consider the ramifications of possible courses of action or interpretation. You will give an economic interpretation to the logic of uncertainty. In the style of Gosset you will supply "*real* error bars" around your best estimates, showing that sampling-based confidence intervals are only one element—perhaps a quite small element—of the discussable error.[16] You will, in other words, draw a dividing line of believable effect size at which some phenomenon should be considered scientifically or humanly important. You will devote your energies to examining the substantive deviations from this minimum oomph.

A tall order? Yes, but it has an honored name. It is called "science." Variable by variable, model by model, it is a difficult change to make, from the nonscience of statistical "significance" to an actual science of oomph. It moves away from the metaphysics of "existence" characteristic of Greek-derived mathematics and philosophy to the calculable magnitudes characteristic of modern science since the seventeenth century. It's hard to do, unlike calculating t-statistics, which is a simpleton's parlor game. But actual science at the frontier is supposed to be difficult. If it wasn't, you wouldn't be at the frontier.

The other problem we are highlighting is what is known as the *fallacy of the transposed conditional*. None of the oomph-lacking tests of Student's *t*, as its inventor told Fisher and Egon Pearson in 1926, are logically speaking tests of hypotheses at all. Fisher is the reason. He erased the Bayesian odds from Gosset's original test of hypotheses, ignored Gosset's and Neyman's and Pearson's insights about the power of the test, and instead calculated the likelihood of observing the given data, assuming a single null hypothesis is true. In some instances of science Fisher's inversion may pose no particular problem. It depends on the question one is asking. But in daily use of Fisher's methods the logic is turned on its head: the sizeless scientists claim to observe the likelihood of a null hypothesis, assuming the data they happen to have in hand are true, the exact reverse of what Fisher's method produces. "If *H*, then *O*" is supposed to affirm "If *O*, then *H*." It doesn't.

If a person is hanged, he will probably die. Therefore, say the sizeless scientists on coming upon a corpse in the street, "He was probably hanged." Something is amiss. The probability of being dead, given that you were hanged, is much higher—much more *statistically* significant, as irrelevant as such a proposition is for finding out exactly why a person is dead—than the probability that you were hanged given that you are dead. This is the fallacy of the transposed conditional. A high likelihood of the sample, supposing the hypothesis is true, is supposed to imply a high probability of the truth of the hypothesis in light of the sample. No. Bringing back Bayes's rule, as sophisticates such as Lindley and Zellner and Good and others have done, is probably a good idea (Bernando 2006). But on our two main points a Bayesian revolution is not necessary, merely efficient. What *is* necessary is to clearly distinguish Gossetian *hypothesis* testing from Fisherian *significance* testing.

In other words, since the 1920s many economic and medical and psychological and forensic scientists have calculated the wrong probability. To take one of by now literally millions of examples, the fallacy of the transposed conditional has in psychometrics led to a gross overestimate of the number of adults afflicted by schizophrenia (Cohen 1994, 999–1000). You can well imagine the need for and relevance of the size-matters/how-much question in all of the sciences.

After Fisher, then, the sizeless sciences neither test nor estimate. They practice a third method of science not easily recognized by a Fisher-only education. The third method—which is a marriage of the sizeless stare of statistical significance to the fallacy of the transposed conditional—we call *testimation,* the ruin of empirical research.

Fisher's testimation has arrived at high places, such as the Supreme Court of the United States. In *Castenda v. Partida,* 430 U.S. 482 (1977), concerning jury discrimination, the court held that "as a general rule for such large samples, if the difference between the expected value and the observed number is greater than two or three standard deviations [that is, if $t > 2.0$ or 3.0 or $p < .05$ or $.01$], then the hypothesis would be suspect to a social scientist, 430 U.S. at 496 n.17."[17] This is mistaken. By God's grace the estimate of jury discrimination in question may be good or bad. But its sheer statistical significance is no evidence one way or the other. If the variable doesn't fit / You may not have to acquit. It depends on the oomph, the expected loss of sticking to the null. A suspect and his jury deserve to get from the expert a clear statement about the warrants of maintaining one hypothesis over another given the truth of the observed data. Today they get instead a statement about the warrants of maintaining the observed data, assuming the truth of a maintained hypothesis (the null hypothesis of, say, "innocent"). The law and all such work—that is to say, most of the statistical work in economics, psychology, medicine, and the rest since the 1920s and especially since the 1980s and the coming of personal computers—has to be done over again.

But the Standard Error Is Tempting: Gosset Knew

Gosset himself never believed "significance" was a substitute for finding out How Much. He was one of nature's economists and was required to act as a profit center at Guinness, where he was an apprentice brewer in the Experimental Division (1899–1906), later head experimental brewer (1907–35), and for the rest of his short life head brewer of Guinness in both Dublin and the newly established brewery in London, Park Royal (1935–37).[18] Gosset's field experiments with barley varieties in Ireland and England yielded small samples. Agricultural experiments are expensive, he knew firsthand, so large sample sizes, on which statistical theory since the eighteenth century had been based, were not profitably relevant. How then could he distinguish the mean difference between, say, two barley yields with $N = 4$? Even with small samples the evidence for or against a null hypothesis of no difference was, Gosset told Karl Pearson in an important letter of 1905, a matter of net "pecuniary advantage."

My original question and its modified form. When I first reported on the subject [of "The Application of the 'Law of Error' to the Work of the

Brewery"], I thought that perhaps there might be some degree of probability which is conventionally treated as sufficient in such work as ours and I advised that some outside authority in mathematics [such as Karl Pearson] should be consulted as to what certainty is required to aim at in large scale work. However it would appear that in such work as ours the degree of certainty to be aimed at must depend on the *pecuniary advantage to be gained by following the result of the experiment, compared with the increased cost of the new method, if any, and the cost of each experiment.* This is one of the points on which I should like advice. (Gosset, ca. April 1905, in E. Pearson 1939, 215–16; italics supplied)

Pearson didn't understand the advice Gosset was requesting and certainly never realized that Gosset was the one who was giving the advice. The great man of large samples never did grasp Gosset's point—though wisely he agreed to publish "Student's" papers.

Gosset seems never to have tired of teaching it. Twenty-one years after his letter of 1905 he responded to a query by Egon Pearson (1895–80), the eldest son of the great Karl, who, unlike Pearson *père*, definitely did grasp the point. In his response Gosset improved on his already sound definition of *substantive* significance. To net pecuniary value he added that before she can conclude anything decisive about any particular hypothesis the statistician must account for the expected "loss" [measured in beer bitterness or lives or jobs or pounds sterling] relative to some "alternative hypothesis."[19] Gosset explained to Pearson *fils* that the confidence we place on or against a hypothesis depends entirely on the confidence and real world relevance we put on some *other* hypothesis, possibly more relevant.

Gosset's letters to the two Pearsons, and his twenty-one published articles, we have noted, are essentially unknown to users of statistics and especially to economists. Yet Gosset was proposing, and using in his own work at Guinness, a characteristically economic way of looking at the acquisition of knowledge. He focused on the opportunity cost, the value of the sacrifice incurred by choosing one of the competing hypotheses. It became the way of Neyman-Pearson and Wald and Savage, though crushed in practice by Fisher's forceful, antieconomic campaign. Fisher, by contrast, looked for some absolute qualitative essence in the aristocratic style of philosophy or theology or even of economics itself in the age before the idea of opportunity cost was made clear.[20] After the Gosset letters of May 1926 Egon Pearson and Jerzy Neyman began a series of famous theoretical papers establishing beyond cavil that Fisher was wrong and Gosset right. To no avail.

"*Pecuniary* advantage," Gosset's invention, is not the unique currency of a regression coefficient or the difference between two means. Gosset, a man of sense and compassion, well understood this—though even today in cancer epidemiology the merely financial expense has a partial claim, considering the alternative employment of the money in saving other lives. Gosset himself conducted original studies on genetics, yeast, barley, hops, the water of the Thames, and, years later, as we have noted, the nutritional advantage to children of drinking raw milk (Student 1931a). He focused always on the substantive meaning—the *chemical* and *biological* significance of his coefficients. If you yourself deal in medicine or psychiatry or experimental psychology, Gosset and we would recommend that you focus on *clinical* significance. If you deal in complete life forms, *environmental* or *ecological* significance. If you deal in autopsies or crime or drugs, *forensic* or *psychopharmacological* significance. And so forth. In short, Gosset's rule is: in any science, attend to oomph. An arbitrary and Fisherian notion of "statistical" significance should never occupy the center of scientific judgment.

A great yet essentially unknown scientist, Gosset quietly invented, among other things, the definition of economic significance, the statistical "design of experiments," the table of *t*, the *t*-test, and even the ideas of "alternative hypotheses," "power," and "loss" (Ziliak 2008a). He did not consider economic calculation or power or loss as mere add-ons or optional accoutrements in case "you have time" or "are curious" after grinding out significant *t*'s and high R^2. Gosset shrugged at a merely statistical significance found in the single sample on offer. Late in life he wrote a letter to Egon, who had recently succeeded his father as editor of *Biometrika*.[21] Gosset was working on experiments with another old friend, Edwin S. Beaven (1857–1941), a pioneer in agricultural experiments and the world's leading authority on barley and malt.[22] Gosset wished to publish in *Biometrika* the results of their experiments together with his own latest thinking about the role of "significance" in the design of experiments. "The important thing in such," Gosset wrote, "is to have a low real error, not to have a 'significant' result at a particular station [as Fisher sought]. *The latter*"—that is, a merely statistical significance defined in Fisher's way, he told the new editor of *Biometrika*—"*seems to me to be nearly valueless in itself.*"[23]

Gosset was prophetic against the mechanization of statistical instruments, too, including even calculating machines. He computed the table of *t* with a mechanical calculator, "Baby Triumphator," motored by a turn-crank and his own strong arm. But he had used electric machines, too, at the brewery and later at Fisher's office, and felt the intellectual difference. The same dangerous ease of calculation has brought statistical significance to a peak in our own Early Computer Age. In Gosset's 1905 report on "The Pearson Co-efficient of Correlation," he warned his fellow brewers that "the better the instrument the greater the danger of using it unintelligently. . . . Statistical examination in each case may help much, but no statistical methods will ever replace thought as a way of avoiding pitfalls."[24] Statistical instruments, such as the *t*-test, the correlation coefficient, and Intel Inside/Celeron, will not replace thought. Precisely. Some years later the poet and Latin textual critic A. E. Housman complained about the replacing of thought with thoughtlessly mechanical rules (for example, "Honor the existing texts even if they yield nonsense") in an article entitled, with heavy sarcasm, "The Application of Thought to Textual Criticism" (Housman 1922 [1961]; see McCloskey 1985 [1998], 72–73). Gosset and we would like to see the application of thought to statistical methods.

But Gosset was not as forceful as Housman or Fisher. He had, a friend of his schooldays said, "an immovable foundation of niceness." He had the virtue of scientific and personal humility, so often misunderstood in post-romantic thought as self-abnegation (McCloskey 2006). He worked "not for the making of personal reputation, but because he felt a job wanted doing and was therefore worth doing well" (E. Pearson 1939, 249). In the rough and tumble politics of the academy, and therefore in business and agriculture and law, a humble brewer of Guinness lost out to a very forceful eugenicist.

As Fisher himself said, "The History of Science has suffered greatly from the use by teachers of second-hand material. . . . A first-hand study is always instructive, and often . . . full of surprises" (quoted in Mendel 1955, 6). Unless you understand the first-hand history you are going to continue thinking—as you do if you are a sizeless scientist and as we once did ourselves—that there must be *some* argument for the 5 percent philosophy. You will suppose that Fisher's way could not be so gravely mistaken—could it? Surely, you will think, a Fisherian disciple of the intellectual quality of Harold Hotelling could not have been confused on the matter of statistical significance. Surely Ziliak and McCloskey and the critics of the technique since the 1880s such as Edgeworth and Gosset and Jeffreys and Deming and Savage and Kruskal and Zellner must have it wrong.

But when you see how Fisher and his immediate followers achieved their sad victory we think you will change your mind. As Friedrich A. Hayek wrote in 1952:

> The paradoxical aspect of it, however, is . . . that those who by the sci-entistic prejudice are led to approach social phenomena in this manner [e.g., accept the phenomena if statistically significant, otherwise reject] are induced, by their very anxiety to avoid all merely subjective elements and to confine themselves to "objective facts," to commit the mistake they are most anxious to avoid, namely that of treating as facts what are no more than vague popular theories. They thus become, when they least expect it, the victims of . . . [what Whitehead called] the "fallacy of mis-placed concreteness." (54)

We think the now sizeless scientists can fulfill the promise of the nine-teenth-century quantitative revolution and become rationally anthropo-metric, biometric, cliometric, econometric, psychometric, sociometric, technometric, and pharmacogenomic. But they will need to get back to questions of *how much* and *who cares?* They will have to shift attention away from their α-selves and toward their β-selves. Statistical scientists share a common intellectual descent with β-Gosset. But most do not re-alize that α-Fisher's methods are a mutation because in this instance of the history of science it was the wiser teacher who was spurned by a greedy apprentice. Vague popular theories inherited from the flawed Fisher, such as that *t*-statistics are objective evidence of the existence of an effect, or that the R^2 possesses a substance-independent scale on which a model is said to be significant or not, or—to say it more generally—that fit is the same thing as importance, will have to go.

Dieting "Significance" and the
Case of Vioxx

The rationale for the 5% "accept-reject syndrome" which afflicts econometrics and other areas requires immediate attention.
                                    ARNOLD ZELLNER 1984, 277

The harm from the common misinterpretation of $p = 0.05$ *as an error probability is apparent.*
                                    JAMES O. BERGER 2003, 4

PRECISION IS NICE BUT OOMPH IS THE BOMB

Suppose you want to help your mother lose weight and are considering two diet pills with identical prices and side effects. You are determined to choose one of the two pills for her.

The first pill, named Oomph, will on average take off twenty pounds. But it is very uncertain in its effects—at plus or minus ten pounds (you can if you wish take "plus or minus" here to signify technically "two standard errors around the mean"). Oomph gives a big effect, you see, but with a high variance.

Alternatively the pill Precision will take off five pounds on average. But it is much more certain in its effects. Choosing Precision entails a probable error of plus or minus a mere one-half pound. Pill Precision is estimated, in other words, much more *precisely* than is Oomph, at any rate in view of the sampling schemes that measured the amount of variation in each.

So which pill for Mother, whose goal is to lose weight?

The problem we are describing is that the sizeless sciences—from agronomy to zoology—choose Precision over Oomph every time.

Being precise is not, we repeat, a bad thing. Statistical significance at some arbitrary level, the favored instrument of precision lovers, reports on a particular sort of "signal-to-noise ratio," the ratio of the music you can hear clearly relative to the static interference. Clear signals are nice, especially so in the rare cases in which the noise of *small samples* and not of misspecification or other "real" errors (as Gosset put it) is your chief problem. A high signal-to-noise ratio in the matter of random samples is helpful if your biggest problem is that your sample is too small, though the clarity of the signal itself is a radically incomplete criterion for making a rational decision.

The signal-to-noise ratio is calculated by dividing a measure of what one wants—the sound of a Miles Davis number, the losing of body fat, the impact of the interest rate on capital investment—by a measure of the uncertainty of the signal such as the variability caused by static interference on the radio or the random variation from a smallish sample. In diet pill terms the noise—the uncertainty of the signal, the variability—is the random effects, such as the way one person reacts to the pill by contrast with the way another person does or the way one unit of capital input interacts with the financial sector compared with some other. In formal hypothesis-testing terms, the signal—the observed effect—is typically compared to a "null hypothesis," an alternative belief. The null hypothesis is a belief used to test against the data on hand, allowing one to find a difference from it if there really is one.

In the weight loss example one can choose the null hypothesis to be a literal zero effect, which is a very common choice of a null. That is, the average weight loss afforded by each diet pill is being tested against the null hypothesis, or alternative belief, that the pill in question will not take any weight at all off Mom. The formula for the signal-to-noise ratio is:

$$\frac{\text{Observed Effect—Hypothesized Null Effect}}{\text{Variation of Observed Effect}}$$

Plugging in the numbers from the example yields for pill Oomph $(20 - 0)/10 = 2$ and for pill Precision $(5 - 0)/0.5. = 10$. In other words, the signal-to-noise ratio of pill Oomph is 2 to 1 and of pill Precision 10 to 1. Precision, we find, gives a much *clearer* signal—five times clearer.

All right, then, once more: which pill for Mother? Recall: the pills are identical in every other way, including price and side effects. "Well," say our significance-testing, sizeless scientific colleagues, "the pill with the highest signal-to-noise ratio is Precision. Precision is what scientists want and what the people, such as your mother, need. So, of course, choose Precision."

But Precision is obviously the wrong choice. Wrong for Mother's weight management program and wrong for the many other victims of the sizeless scientist. The sizeless scientist decides whether something is important or not—she decides "*whether* there *exists* an effect," as she puts it—by looking not at the something's oomph but at *how precisely it is estimated*. Diet pill Oomph is potent, she admits. But, after all, it is very imprecise, promising to shed anything from 10 to 30 pounds. Diet pill Precision will, by contrast, shed only 4.5 to 5.5 pounds, she concedes, but, goodness, it is very *precise*—in Fisher's terms, very *statistically* significant. From 1925 to 1962, Ronald A. Fisher instructed scientists in many fields to choose Precision over Oomph every time. Now they do.

Common sense, like Gosset himself, would of course recommend Oomph. Mom wants to lose *weight,* not gain precision. Mom cares about the spread around her waist. She cares little—or not at all—for the spread around the average of an imaginary, infinitely repeated, random sample. The minimax solution (to pick one type of loss function) is obvious: in all states of the world, Oomph dominates Precision. Oomph wins. Choosing the inferior pill, that is, pill Precision, instead maximizes failure—the failure to lose up to an additional 25.5 (30 −4.5) pounds. You should have picked Oomph.

Statistical significance, or sampling precision, says nothing about the oomph of a variable or model. Yet scientists in economics and medicine and the other statistical fields are deciding about oomph on the basis of this one kind of precision. A lottery is a lottery is a lottery, they seem to be saying. A pile of hay is a pile of hay; a mustard packet is a child.

The attention lavished on the signal-to-noise ratio is difficult to fathom, even for acoustical purists such as the noted violinist Stefan Hersh. "Even *I* get the point about the phoniness of statistical significance," he said to Ziliak one day over lunch. It seems to be hard for scientists trained in Fisherian methods to see how bizarre the methods in fact are and increasingly harder the better trained in Fisherian methods they are.

The level of significance, precision so defined, says what? That "one in a hundred times in samples like this one, if random, the signals will be confused." Or "Nine times out of ten, if the problem is a sampling problem, the data will line up *this* way relative to the assumed hypothesis without specifying how *important* the deviations or signal confusions are."

Logically speaking, a measurement of sampling precision can't possibly be the end of the inquiry. In the sizeless sciences, from economics to medicine, though, it is. If a result is "precise" in the narrow sense of sampling, then it is hailed as "significant."

Rarely do the sizeless scientists speak in Neyman's sampling terms about confidence intervals or in Gosset's non-sampling terms about real "error bars" (Student 1927). Even more rarely do they speak of the relevant range of effects in the manner of Leamer's (1982) "extreme bounds analysis." And still more rarely do they attend to all the different kinds of errors, errors more dangerous, Gosset insisted, than mere error from sampling—which is merely the easiest error to know and to control. They focus and stare fixedly at tests on the single-point percentage of red balls and white balls drawn hypothetically repeatedly and independently from an urn of nature. (Fisherians do not literally conduct repeated experiments. The brewer did.) But the test of "significance" defined this way, a number—a single point in a distribution—without a scale on which to judge its relevance, says almost nothing. It says nothing at all about what people want unless they want only insurance against a particular kind of sampling error—Type I error, the error of undue skepticism—along a scale on which every red ball or white ball has the same impact on life and judgment.

A century and a half ago Charles Darwin said he had "no Faith in anything short of actual Measurement and the Rule of Three," by which he appeared to mean the peak of arithmetical accomplishment in a nineteenth-century gentleman, solving for x in "6 is to 3 as 9 is to x." Some decades later, in the early 1900s, Karl Pearson shifted the meaning of the Rule of Three—"take 3σ [three standard deviations] as definitely significant"—and claimed it for his new journal of significance testing, *Biometrika*.[1] Even Darwin late in life seems to have fallen into the confusion. Francis Galton (1822–1911), Darwin's first cousin, mailed Darwin a variety of plants. Darwin had been thinking about point estimates on the heights of self- and cross-fertilized plants that depart three "probable errors" or more from the assumed hypothesis, a difference in height significant at about the 1 percent level.

But the gentlemanly faith in the New Rule of Three was misplaced. A statistically significant difference at the 1 percent level (an estimate departing three or more standard deviations from what after Fisher we call the null) may for purposes of botanical or evolutionary significance be of *zero* importance (cf. Fisher 1935, 27–41). That is, some cause of natural selection may have a high probability of replicability in additional samples but

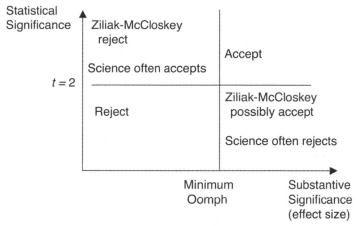

Fig. 1.1. Minimum oomph is what you're looking for or should. (Adapted from figure 1 in Erik Thorbecke, "Economic and Statistical Significance: Comments on 'Size Matters,'" *Journal of Socio-Economics* 33 [5, 2004]: 573. Copyright © Erik Thorbecke 2004, with permission from Elsevier Press.)

be trivial. Yet, on the other hand, a cause may have a low probability of replicability but be important. This is what we mean when we say that a test of significance is neither necessary nor sufficient for a finding of importance. In significance testing the substantive question of what matters and how much has been translated into a 0 to 1.0 probability, regardless of the nature of the substance, probabilistically measured.

After Fisher, the loss function intuited by Gosset has been mislaid. It has been mislaid by scientists wandering our academic hallways transfixed in a sizeless stare. That economists have lost it is particularly baffling. Economists would call the missing value of oomph the "reservation price" of a possible course of action, the opportunity cost at the margin of individual-level or groupwise decision. Without it our actual measurements—our economic decisions—come up short (fig. 1.1). As W. Edwards Deming put it, "Statistical 'significance' by itself is not a rational basis for action" (1938, 30).

Yet excellent publishing scientists in the sizeless sciences talk as though they think otherwise. They talk as though establishing the statistical significance of a number in the Fisherian sense is the *same thing* as establishing the significance of the number in the common sense. Here, for example, is a sentence from an article in economic science coauthored by a scientist we regard as among the best of his generation, Gary Becker (b. 1930), a Nobel laureate of 1992. Becker's article was published in a leading journal in 1994: "The *absolute* t *ratio* [the signal-to-noise ratio, using Student's *t*] associated with the coefficients of this variable is 5.06 in model (i), 5.54 in model (ii), and 6.45 in model (iii). . . . These results suggest [because Student's *t* exceeds 2.0] that decisions about current consumption *depend* on future price" (Becker, Grossman, and Murphy 1994, 404; italics supplied). Notice the rhetoric of depend/not-depend, exist/not-exist, whether/ not, and significant/insignificant even from such a splendid economic scientist as Becker. He has confused a measurement of sampling precision—that is, the size of the *t* statistics—with a quantitative/behavioral demonstration—that is, the size of the coefficients. Something is wrong.

"Significance" and Merck

Merck was in 2005 the third-largest drug manufacturer in the United States. Its painkiller Vioxx was first distributed in the United States in 1999 and by 2003 had been marketed in over eighty countries. At its peak in 2003 Vioxx (also known as Ceoxx) brought in some $2.5 billion. In that year a seventy-three-year-old woman died suddenly of a heart attack while taking as directed her prescribed Vioxx pills. Anticipating a lawsuit the senior scientists and company officials at Merck, newspaper accounts have said, huddled over the statistical significance of the original clinical trial.

From what an outsider can infer, the report of the clinical trial appears to have been fudged. Data that made Vioxx look bad were allegedly simply omitted from the report. A rheumatologist at the University of Arizona and lead author of the 2003 Vioxx study, Jeffrey Lisse, admitted later that not he but Merck "actually wrote the report." Perhaps there is some explanation of the Vioxx study consistent with a more reputable activity than data fudging. We don't know.

"Data fudging and significance testing are not the same," you will say. "Most of us do *not* commit fraud." True. But listen.

The clinical trial was conducted in 2000, and the findings were published three years later in the *Annals of Internal Medicine* (Lisse et al. 2003). The scientific article reported that "five [note the number, five] patients taking Vioxx had suffered heart attacks during the trial, compared with one [note the number, one] taking naproxen [the generic drug, such as Aleve, given to a control group], *a difference that did not reach statistical significance.*"[2] The signal-to-noise ratio did not rise to 1.96, the 5 percent level of significance that the *Annals of Internal Medicine* uses as a strict line of demarcation, discriminating the "significant" from the insignificant, the scientific from the nonscientific, in Fisher's and today's conventional way of thinking.

Therefore, Merck claimed, given the lack of statistical significance at the 5 percent level, there was *no* difference in the *effects* of the two pills. No difference in oomph on the human heart, they said, despite a Vioxx disadvantage of about 5 to 1. Then the alleged fraud: the published article neglected to mention that in the same clinical trial *three additional takers of Vioxx,* including the seventy-three-year-old woman whose survivors brought the problem to public attention, suffered heart attacks. Eight, in fact, suffered or died in the clinical trial, not five. It appears that the scientists, or the Merck employees who wrote the report, simply dropped the three observations.

Why? Why did they drop the three? We do not know for sure. The courts are deciding. But an outsider could be forgiven for inferring that they dropped the three observations *in order to get an amount of statistical significance low enough to claim*—illogically, but this is the usual procedure—*a zero effect.* That's the pseudo-qualitative problem created by the backward logic of Fisher's method. Statistical significance, as the authors of the Vioxx study were well aware, is used as an on-off switch for establishing scientific credibility. No significance, no risk to the heart. That appears to have been their logic.

Fisher would not have approved of data fudging. But it was he who developed and legislated the on-off switch that the Vioxx scientists and the *Annals* (and, to repeat, many courts themselves) mechanically indulged. In this case, as in many others, the reasoning is that if you can keep your sample small enough—by dropping parts of it, for example, especially, as in this apparently fraudulent case, the unfavorable results—you can claim *in*significance and continue marketing. In the published article on Vioxx you can see that the authors believed they were testing, with that magic formula, whether an effect existed. "The Fisher exact test," they wrote in typical sizeless scientific fashion, and in apparent ignorance of the scientific values of Gosset, "was used to compare incidence of confirmed perforations, ulcers, bleeding, thrombotic events, and cardiovascular events. . . . All statistical tests . . . were performed at an α level of 0.05" (Lisse et al. 2003, 541).

If the Merck scientists could get the number of heart attacks down to five, you see, they could claim to other sizeless scientists that the harmful effect wasn't there, didn't exist, had no oomph, was in fact zero. The damage was actually naproxen takers one victim, Vioxx takers *eight* victims, not five. Other things equal, the relative toll of Vioxx to naproxen was 8 to 1, leaning strongly against Vioxx. And with the sample size the scientists had the true eight heart attacks were in fact statistically significant even by the 5 percent Fisher criterion—good enough, that is, by their own standard of sampling precision, to be counted as a scientific "finding" in the *Annals*. But Merck didn't want to find that its Vioxx was dangerous. So it pretended that the deaths were insignificant.

In a scientific culture depending on a crude version of precision and the sizeless stare, "significance" was, sociologically speaking, Merck's problem. Merck wanted the unfavorable results to be statistically *in*significantly different from a zero effect so that it could claim no effect. It misunderstood the significance of significance. That was not, of course, the sin itself. Dropping the three observations was the sin, if in truth it happened. But, as Roman Catholic theologians put it, the *occasion* for sin appears to have been the Fisherian rhetoric of 5 percent significance.

At five *or* eight in the "failure" class the sample size, you might say, must have been too small to make the judgment: with such small numbers one cannot tell *what* is important. But that's not right, since what matters is the total sample size, not the rare heart attacks, a sample size that was anyway large enough to satisfy the editors of the journal. And anyway, small samples can show important effects. World War I happened only once ($N = 1$), yet it was significant. You were born only once ($N = 1$), yet you have loved and lost. One California man (insignificant at the .05 level) threw a woman's dog into oncoming traffic ($N = 1$), and the state responded by toughening "road rage" laws. Gosset himself invented Student's t with a sample of bulk barley of size $N = 2$ (Student 1908a, 23). A small sample, we repeat, is rarely the big scientific problem. Interpretation is.

Gosset would have rejected the interpretation of the Vioxx scientists and their "insignificant" 5-to-1 ratio of heart attacks. Statistical significance or its lack at an arbitrarily high or low level is not the issue, Gosset always said. The 5 percent philosophy invented by Fisher and enforced by the *Annals of Internal Medicine,* Gosset would say, was part of the problem, not the solution. "What the odds should be," Gosset wrote in 1904, "depends: (1) On the degree of accuracy which the nature of the experiment allows, and (2) On the importance of the issues at stake."[3] Merck wanted "importance of the issues at stake" to mean "odds of an absolute criterion, $p < 0.05$, regardless of the importance of the loss or gain from the drug." Therefore Merck said that "there were too few end points to allow . . . authoritative conclusions about the relative effects . . . on cardiovascular events" (Lisse et al. 2003, 545).

Widows and widowers and sound-thinking scientists are on Gosset's side. But internally at Merck it was tough for scientists to be. It appears from newspaper accounts that a Dr. Edward Scolnick, a top research scientist at Merck from 1985 to 2002, was silenced internally for saying in a company e-mail that the "benefits and risks" of Vioxx have not been "fairly" considered, that statistical significance was being treated in an un-Gossetian way.

Merck took Vioxx off the market. But it is in trouble and faces many trials in a nonstatistical sense of the word (more than 4,200 suits had been filed as of August 20, 2005). The lawsuits over Vioxx are going to force Merck's lawyers, alas, to defend the Fisherian misuse of statistical significance. If an attorney on the anti-Merck side can grasp the argument we are making here and persuade a judge or jury that sizeless science is nonscience, she will make herself and her clients very rich and make new and better law and encourage new and better science.

The Whale of Significance

Our colleagues in the sizeless sciences get very upset by our Vioxx story. But they don't offer persuasive reasons for 5 percent science. Unreasoning anger is a quite common reaction to challenges to the Fisherian orthodoxy. We implore our colleagues not to use their anger to dodge the main point. Tell us, please, what the *arguments* for Fisherian procedures are. Don't merely get angry at our style or our presumption or our appeals to a beer brewer. Tell us where we go wrong.

Another story. The Japanese government in June 2005 increased the limit on the number of whales that may be annually killed in Antarctica—from around 440 annually to over 1,000 annually. Deputy Commissioner Akira Nakamae explained why: "We will implement JARPA-2 [the plan for the higher killing] according to the schedule, because the sample size is determined in order to get statistically significant results" (Black 2005). The Japanese hunt the whales, they claim, in order to collect scientific data on them. That and whale steaks. The commissioner is right: increasing sample size, other things equal, does increase the statistical significance of the result.[4] It is, after all, a mathematical fact that statistical significance increases, other things equal, as sample size increases. Thus the theoretical standard error of JARPA-2, $s/\sqrt{(440 + 560)}$ [given for example the simple mean formula], yields more sampling precision than the standard error of JARPA-1, $s/\sqrt{(440)}$. In fact, it raises the significance level to Fisher's 5 percent cutoff. So the Japanese government has a found a formula for killing more whales, annually some 560 additional victims, under the cover of getting the conventional level of Fisherian statistical significance for their "scientific" studies.

Around the same time that significance testing was sinking deeply into the life and human sciences, Jean-Paul Sartre noted a personality type. "There are people who are attracted by the durability of a stone. They wish to be massive and impenetrable; they wish not to change." "Where, indeed," Sartre asked, "would change take them? . . . What frightens them is not the content of truth, of which they have no conception, but the form itself of truth, that thing of indefinite approximation"(1948, 18). Sartre could have been talking about the psychological makeup of the most rigid of the significance testers.[5]

Significance unfortunately is a useful means toward personal ends in the advance of science—status and widely distributed publications, a big laboratory, a staff of research assistants, a reduction in teaching load, a better salary, the finer wines of Bordeaux. Precision, knowledge, and control. In a narrow and cynical sense statistical significance is the way to achieve these. Design experiment. Then calculate statistical significance. Publish articles showing "significant" results. Enjoy promotion.

But it is not science, and it will not last.

2

The Sizeless Stare of Statistical Significance

We hear with horror of the loss of 400 men on board the Birken-
head by carelessness at sea; but what should we feel, if we were told
that 1,100 men are annually doomed to death in our Army at home
by causes which might be prevented?

FLORENCE NIGHTINGALE 1858, 249

Do we exaggerate? Sadly, no. Consider the journals of the sizeless sciences.

SIGNIFYING LITTLE

The very word *significance,* emphasizing its statistical meaning, is often used prominently in an advertisement, especially for pharmaceuticals. Xanax is a product of the Upjohn Company designed to alleviate clinical anxiety. In May 1983 Upjohn ran a two-page spread for Xanax in the *Journal of Clinical Psychiatry* typical of the way *significance* has been used for decades. The science justifying the pill is summarized prominently.

> In double-blind, placebo-controlled clinical trials in 976 patients with moderate to severe clinical anxiety, therapy with Xanax was compared to diazepam (Valium). Patients treated with Xanax *had a significantly lower incidence of drowsiness when compared directly to diazepam* therapy (Valium) in a 976-patient, placebo-controlled, multi-center study. . . . Special analysis of 692 anxious patients with a *significant* depressed mood item score showed that treatment with Xanax was *significantly* better than placebo in decreasing depressed mood score. (*Journal of Clinical Psychiatry* 44 [7, July 1983]: 255–56; first italics in original; others supplied)

This, of course, is advertising. But did the original scientific article back up the advertising with scientific magnitudes? No. The scientific article, published in 1981, gave no indication of the *amount* of depression reduction, however it might be defined. The author's only standard of "better" was statistical significance (Cohn 1981, 349–50). In both the scientific article and the advertisement no evidence is put forward on *how much* drowsiness was avoided—measured by, say, hours of sleep reduction. Nothing.

But surely this is *bad* science. No, probably it is not, not by the standards internal to the sizeless sciences. The *Journal of Clinical Psychiatry* does not publish bad science—or so, at least in humility, a reader from the outside, looking at conventional journal rankings, must suppose. But she can still see how much these otherwise good scientists depend on "significance." You can, for example.

Consider this piece of science from a leading journal of economics. Skip over the economic jargon and note instead the significance rhetoric.

> The coefficient is *significant at the 99 percent confidence level.* Neither the current money shock nor all 12 coefficients as a group are *significantly different from zero.* The coefficient on *c* is *negative and significant and the distributed lag on c is significant as well.* In column (2) we report a regression which *omits the insignificant* lags on money shocks. The *c* distributed lag is *now significant at the 1 percent confidence level.* . . . We interpret these results as indicating that the primary factor determining cyclical variations in the probability of leaving unemployment is probably heterogeneity. . . . [However,] money shocks have no *significant* impact. (*American Economic Review*, September 1985, 630; italics supplied)

Money is complicated, and these fellows are experts in its complexity. But forget about money, unemployment, and "heterogeneity," whatever that may mean. Concentrate on what you can spot even as an outsider to economic science, namely, the rhetoric of Whether and How Much. Note the phrases "significantly different from zero" and "now significant at the 1 percent confidence level" and five other instances of significance talk. Ask: do the authors exhibit evidence of the *sizes* of the alleged effects between money supply and unemployment (however those might be defined) or of risk of loss in jobs or inflation (however those might be measured)? Do they ask How Much? No.

Do they instead report the claimed presence or absence of noise—a mere "significance test"? Yes. Have they translated substantive oomph into a probability space in which they claim they do not need to think about how big is big? Yes. "Money shocks have no *significant* impact," by which they mean "insignificant at the 1 percent level assuming that the only source of error is sampling error along a sizeless scale." They act as if size doesn't matter.

We do not mean to suggest that the authors actually do not care about the economic magnitudes they are talking about. As good scientists they certainly do, somewhere buried deep. But their arguments for this or that number or "impact," we are saying, are based on the irrelevant criterion of sampling precision.

Commonly the sizeless scientists in economics do not even report the impact found, only the existence of an impact. People who really want to know about the economy, and want to do something with the knowledge, will ask questions such as, "If the Federal Reserve Bank decreases the money supply by X percent, how much will society have to pay in the form of higher unemployment"? Existence is not the issue. Number of jobs is the issue or the rate of price inflation or both. The authors here, as even an outsider to economics can see, do not ask the How Much question, at any rate so far as their statistical proofs are concerned. They gaze at the economy with a sizeless stare.

Consider this extract from a 1970 issue of the *New England Journal of Medicine,* another early statistical article on clinical depression. Look, in other words, for the occurrence of precision versus oomph.

> Figure 1 shows that on the seventh day, about 56 hours after the first dose of [tri-iodothyronine] (fifth day), these patients were *significantly* more improved than those given [the placebo] (*p* < 0.01, *Student's* t *test*). *The differences reached the level of statistical significance, by and large,* from the seventh to the sixteenth day and persisted until the final days of the study. . . . The *statistically reliable* benefit that [tri-iodothyronine] patients showed after 56 hours' therapy is compatible. (*New England Journal of Medicine* 282 [19, May 7, 1970]: 1064–66; italics supplied)

Clinical depression, like money and unemployment, is a complicated subject. But you don't have to be a statistical psychiatrist or psychopharmacologist to understand that the scientists here are under the spell of statistical significance. "The differences reached the level of statistical significance, by and large," they say. All sampling precision, no substantive oomph. Unlike the authors of the money and unemployment article, though, the authors of the depression article do at least refer to something called a "*small* dose" of tri-iodothyronine. That's a step toward the measurement and interpretation of a size that matters. A small step.

Consider this extract, again in psychiatry, and again concerning the anxiety-reducing Xanax drug.

> The side effect reported by the greatest number of patients (394), drowsiness, appeared *significantly more frequently* in the [Valium] group than in the [Xanax] group ($p < .05$). None of the other side effects reported were *significantly different* for the three groups. (*Journal of Clinical Psychiatry*, September 1981, 349–50.)

"Significantly more frequently," they report in sizeless style. Equivocation about the meaning of the very word *significant* is a sign that the authors have no idea of actual magnitudes. Untutored common sense should say, quite properly, "Why does that matter? What are the *actual hours* of sleep induced?" Valium users want to know.

Or again, in epidemiological science, as late as 2003:

> Model reduction was terminated when all variables in the model *were significant at $p < 0.05$.* . . . Of the variables evaluated in the bivariable models for association with the odds of a pig being seropositive [i.e., infected with salmonella], . . . [these] were found *significant at $p < 0.25$.* After the stepwise reduction, only one [variable], CATGROWTH, was *significant at the 5% level.* . . . There was *no significant ($p = 0.35$) difference* in the risk of seropositivity [of salmonella] between the pigs that were fed non-pelleted dry and wet rations. (*Epidemiology and Infection* 131 [2003]: 601–2)

The risk of death from salmonella poisoning can be high, as this study of Greek swineherds may or may not show. But you aren't being told. Variables were cut off at the 5 percent level. And of the survivors the authors do not report the oomph of competing risks. One is moved to haiku.

> A variable
> was murdered at 5 percent:
> justice not pursued.

Or take a case in medical science. Find the scientific error.

> *Student's* t *test* was used to calculate the *statistical significance of differences* in the mean concentrations of amino acids between the two groups of phenylketonuric patients and the group of normal subjects. The con-

centrations of most amino acids in the plasma of the untreated, mentally defective patients with phenylketonuria were *significantly lower (p < 0.01)* than those in the normal [intelligence] subjects. (*New England Journal of Medicine* 282 [14, April 2, 1970]: 764; italics supplied)

Poor, neglected Student. Statistical significance will never by itself prove anything about how much the difference is between mentally "normal" and mentally "defective" patients. That is a matter of how much one *cares* about the ability to rotate objects in one's mind or to count backward from ninety-seven by fives. Intelligence is a matter of How Much, What Kind, and For What Purpose. It is not about a sizeless existence.

Existence talk is rife in studies of education. Critical thinking is a priority of good teachers in a democracy, of course. But how much critical thinking and of what kind? Forty teachers of secondary students in Long Island, New York, were randomly surveyed. In the survey they were questioned about their personal "CT-beliefs" (critical-thinking beliefs) and practices. The study was published in 2005 in a high-profile journal of education. Here is how the authors of the study present the statistical evidence.

Selection of prompts and advantage-characteristic items for use in subsequent studies was made according to the following criteria: (a) prompts yielding a statistically significant main effect in multivariate analysis of group membership, (b) items with a strong association to the main effect (i.e., statistically significant univariate *F* values), . . . and (d) distribution across the five secondary subjects (English, languages other than English, mathematics, science, and social studies). . . . Twelve prompts were retained that discriminated between groups at the conservative alpha level of .0025. . . . On high CT-prompts for high-advantage learners, the difference between CT-inclined teachers . . . and CT-adverse teachers . . . was statistically significant. (*Educational and Psychological Measurement* 65 [2005 (1)]: 159–60)

This is not the critical thinking of, let's say, John Dewey or Paulo Freire. The intentions were good. The authors thought to compare critical-thinking-inclined teachers with other kinds of teachers. But the authors work near the bottom of Bloom's Taxonomy of educational achievement, repeating words and figures they significantly misunderstand.

Spot the oomph, if it's there, in forensic science, in a report of 2003.

[We aim to] assess *significant* bivariable associations of the independent measures and increased condom use for respondents. . . . Variables were . . . retained if they were *statistically (p < .05) or marginally*

(.10 > p > .05) significant. . . . Table 1 shows that increased condom use with a steady partner was significantly associated (p < .05) with drug using status, intervention condition received, age at first use of the male condom, and prior use of the male condom as a means of birth control. . . . Increased condom use with casual paying partner(s) was significantly (p < .05) associated with crack cocaine use intensity. (Journal of Drug Issues 33 [2003 (1)]: 10–11 [lead article]; italics supplied)

The authors make a claim that is usual in the sizeless sciences. The investigator, they say, should "retain" variables—that is, use them as inputs for further statistical calculation, "if they [are] statistically (p < .05) or marginally (.10 > p > .05) significant." No oomph. They learned to think this way from a student of a student of a student of a faulty old book by Ronald A. Fisher. Edgeworth, Gosset, Borel, and a long line of other sensible users of statistics, by contrast, do not think Fisher's routine is socially useful or minimally logical.

Again, in animal science:

The object of these experiments was to study the effects of a tranquilizer on the inducement of ewes to accept orphan lambs by relieving the nervous stress brought about by such transfers. . . . [The "orphans" were healthy lambs separated at birth by the research workers from their biological mothers.] The hypothesis for calculation of the ["Fisher"] exact probability was based upon the assumption that if there were no treatment effect, injected ewes would react similarly to control ewes. Five of six injected ewes accepted and raised their orphan lambs to a weaning weight of 19.14 kg (p < 0.05). One lamb was allowed to nurse following injection but was observed to be dead at 08.00 hours the following day. One control ewe accepted her orphan lamb after transfer and raised it until weaning. The other five lambs [assigned to nontranquilized foster mothers] were rejected, and three of these died at 24 hrs after transfer and two died about 48 hrs after transfer. Death presumably was caused by injury imposed by butting of ewe and/or lack of milk for lamb. . . . When data of experiments 1 and 2 were combined, the effect of 2 ml of perphenazine was highly significant (p < 0.01). (Animal Behaviour 19 [1971 (1)]: 75–79)

The scientists knew in advance, of course, that the untreated mothers—the ewes that were not given the tranquilizer—would either kill outright or neglect to nurse the orphaned lambs. Prior to the experiment the "highly significant" effect was already known to any shepherd. But the ceremony of statistical significance needed to be performed and the torturing of lambs carried out, to assure the scientists of their scientific standing. Traveling down the narrow road to the south of science, haiku pour out of us:

> Here's a Scientist
> who hangs the world on a *t*—
> of Significance.

One would think that in the scientific study of business the focus would be on the bottom line, not on an absolute standard of *statistical* significance. But here is a representative extract from a leading journal of management science.

> Our first hypothesis suggested that visionary leadership *was related to* higher levels of internal and external cooperation. We used two measures to represent internal and external cooperation, quality philosophy and supplier cooperation. Top management team involvement, our measure of visionary leadership, *was significantly related to* both quality philosophy ($t = 10.80$, $p < .001$) and supplier involvement ($t = 7.59$, $p < .001$). Therefore, Hypothesis 1 is supported. (*Decision Sciences* 35 [3, summer 2004]: 407; italics supplied)

"Quality philosophy" is the tip-off. Ironically, the authors were trying to "test" the business philosophy of the great statistician W. Edwards Deming, a founding father of quality control, a postwar hero of Japanese economic growth, and a learned and vehement opponent of Fisherian significance testing.[1] Hypothesis 1 (*H*) is *not* "supported" by the observations (*O*). By Deming's standards the hypothesis wasn't even tested. You are not told how big the effects of "visionary leadership" are. And the fallacy of the transposed conditional runs rampant here as it does in all the cases we've mentioned. The Fisher test "If *H*, then *O*" is mysteriously transformed (the authors never say how) into "If *O*, then *H*." The experts in business management think, in other words, that the probability of making a profit, given that you are a visionary leader, is the same as the probability of being a visionary leader, given that you made a profit. Something again is wrong, a broken link in the chain of inference. Deming knew. From the late 1930s on, his philosophy of business argued that Fisher's "tests"—which the authors of the article in *Decision Sciences* reproduce in orthodox fashion—entail substantively insignificant profit. If you follow Fisher:

You're the business firm
that fits the world to a t—
Insignificant.

You gaze at your clubs
And wish for a hole in one,
Forgetting the ball.

Finally, look at a study from 2003 on the social psychology of recreational drugs. You can see immediately that the study is oomphless. Count the instances of oomphless claims.

> When examined in bivariable analyses, 15 of the 16 temptations-to-use drugs items were found to be associated [i.e., statistically significantly related] with actual drug use. These were: while with friends at a party ($p < .001$), while talking and relaxing ($p < .001$), while with a partner or close friend who is using drugs ($p < .001$), while hanging around the neighborhood ($p < .001$), when happy and celebrating ($p < .001$), when seeing someone using and enjoying drugs ($p < .05$), when waking up and facing a tough day ($p < .001$), when extremely anxious and stressed ($p < .001$), when bored ($p < .001$), when frustrated because things are not going one's way ($p < .001$), when there are arguments in one's family ($p < .05$), when in a place where everyone is using drugs ($p < .001$), when one lets down concerns about one's health ($p < .05$), when really missing the drug habit and everything that goes with it ($p < .010$), and while experiencing withdrawal symptoms ($p < .01$). . . . The only item that was not associated with the amount of drugs women used was "when one realized that stopping drugs was extremely difficult." (*Journal of Drug Issues* 33 ([2003 (1)]: 171–72)

We count sixteen instances of precision-only considerations in this one paragraph—and zero instances of oomph. Its oomph-to-precision ratio is zero percent. The authors believe they are doing serious scientific research, and we suppose that in many ways they are, though we find it strange that they used "bivariable" instead of multiple regression techniques. But their research was funded by a grant from the Centers for Disease Control, with implications for the war on drugs, and surely that certifies their science as serious. Yet their use of statistical significance is clinical. The whole of their world is significant ($p < .05$). They are in the grips of their own addiction to recreational statistics.

Our examples are not a biased selection of the worst. You can test this by taking down the journals yourself in fields such as quantitative economics, general medicine, epidemiology, psychiatry, pharmacology, management science, animal science, education, law, and social psychology and looking for the place—we call it the "crescendo" of the article—at which statistical significance is said to imply the result. Judging by the prestige of the journals and the seriousness of the articles in other ways, the articles we mention here attain the average standard for the leading contributors to the science.

In other words:

(1) The average article in the leading journals of the statistical sciences claims that size doesn't matter;

(2) To put it another way, the average statistical proposition fails to provide the oomph-relevant quantitative information on the phenomena it claims to study;

(3) The average statistical article in the leading journals of science commits the fallacy of the transposed conditional, equating "the probability of the data, given the hypothesis" with "the probability of the hypothesis, given the data";

(4) To put it another way, false hypotheses are being accepted and true hypotheses are being rejected.

We find the results strange. The part of civilization claiming to set empirical standards for science and policy has decided to use illogical instruments and irrelevant empirical standards for science and policy. The quantitative sciences have become pseudo-qualitative and highly illogical at their very center. In journals such as the *New England Journal of Medicine, Journal of Clinical Psychiatry, Annals of Internal Medicine, Animal Behaviour, Educational and Psychological Measurement, Epidemiology and Infection, Decision Sciences*, and *American Economic Review* the oomph, the size of effects, the actual testing of hypotheses, the risks and kinds and sizes of loss, the personal and social meanings of relationships do not seem to matter much to the scientists. An arbitrary level of statistical significance is the only standard in force—regardless of size, of loss, of cost, of ethics, of scientific persuasiveness. That is, regardless of oomph.

3

What the Sizeless Scientists
Say in Defense

*[It] is principally by the aid of [significance testing] that these stud-
ies may be raised to the rank of sciences.*

R. A. FISHER 1925a, 2

You will be uneasy. You will say again, "How can this be?" Statisticians
and statistical scientists sometimes think they disagree with our strictures
on null-hypothesis significance testing (Hoover and Siegler 2008; Mc-
Closkey and Ziliak 2008). They are made unhappy by our radical assertions. When they do not simply become outraged—which is not uncom-
mon, we said, though seldom accompanied by actual argument—they cast
about for a reply.

Some hope they can trip us up on technical details. They observe, for
example, that *both* diet pills in our Mom example achieved a signal-to-
noise ratio of at least 2.0. Therefore, they say, *both* pills are statistically
significant—and so the precision criterion is entirely satisfied. Therefore
the rational scientist will choose the pill Oomph, which dominates the pill
Precision in practical, weight-loss effect. Since the signal-to-noise ratio
meets or exceeds 2.0 for each pill, both pills are, they say, "admissible"
(Savage 1954, 114–16). The objection itself illustrates, by the way, the
lack of interest in oomph. We stipulated that one couldn't take both pills.
The objectors are not listening to the substance of the science.

But it is anyway irrational, we would reply in technical mode, to make
decisions in this sequential fashion, first precision, then oomph, first step
A, then step B, then step C. And even such an irrational procedure is not
what scientists in the sizeless sciences actually do. They routinely pick the

pill with the most precision, not the one with the highest oomph after having established precision. As you can see from the various examples we have given, from economics to medicine, the users routinely neglect oomph entirely, assigning to oomph zero weight. They actually never get to step B or C.

Oomph, a measure of possible or expected loss or gain, is not about large effects "in some absolute sense" (as some of our colleagues have put it to us), whatever that might mean. Oomph is about the clinical or economic or ecological or pharmacological or scientific-rhetorical meaning of the deviation from competing hypotheses—the human, substantive meaning of the effect—large or small. Seeing oomph entails, as Gosset and Wald and Savage and their neglected tradition in statistics require, a loss function.[1] In pharmacology, for example, a scientist concerned with oomph will seek to determine whether a side effect of some drug is small enough, other things equal—small enough in harmful effect, having negligible losses—to keep the drug on the market.[2] Size matters all the way down.

Insignificant Does Not Mean Unimportant

"Somehow the numbers are rigged to get the results you want" is another reply. More technical fixes. No. We can change the numbers in the example, leaning harder against Oomph, if you want, and yet come to the same conclusion: that focusing on Precision alone is mistaken.

Leave the pill Precision at "takes off on average 5 pounds plus or minus 0.5 pounds," but change the statistics for the pill Oomph: suppose it takes off 20 pounds on average but now it is even *more* uncertain in its effects. Before it varied by plus or minus 10 pounds. Now suppose it varies by plus or minus 14 pounds.

The signal-to-noise ratio for pill Precision is still 10 to 1—*very* precise. Pill Oomph, however, is now much less precise, only 1.43 to 1. That's noisy data by Fisher's arbitrary Rule of Two. It yields a very imprecisely estimated average for Oomph's effect, at any rate if your notion of "precision" is "sampling odds alone." At 1.43 the signal-to-noise ratio is a good deal less than 1.96, the *t*-value most commonly used as a cutoff. Above the convention of 5 percent significance (which is the same as $t \leq 1.96$), a signal-to-noise ratio of 1.43 is statistically *in*significant. Unworthy of publication in the leading journals of science.

But, we ask again, which pill for Mother, whose goal in all this is to lose weight, not to gain a publication? Oomph promises to shed between

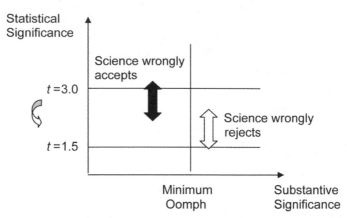

Fig. 3.1. Size matters no matter where you draw the line of precision

6 and 34 pounds while Precision's upper bound is 5.5 pounds. Wild Oomph, despite its lack of precision, is still what Mother wants (fig. 3.1).

The precision-only statistician might say, "Wait a minute. How *much* weight does Mother want to lose? If thirty-four pounds of loss would leave her a skeleton . . ." But again the precision-only statistician is here conceding our case. In asking the how-much question he is valuing substantive significance—weight-losing ability measured in pounds and Mom's well-being—and not a mere "statistical," that is, "probability-of-variance-from-sampling," significance, measured in {0,1} probability. If the pills could not be regulated by dosage or length of treatment, Precision might in fact be the best pill for an already trim Mother who would like to lose only a few pounds for her high school reunion, the better to fit into a size 4 dress. (To which one of the present authors says, "I hate your mother!") The statistician who thinks this way is sensible. His attention is focused on oomph. *How much* weight should Mom lose? *How much* should I save today to secure tomorrow's retirement? *How much* did the outcome of World War I depend on the invention of the tank? *How much* should we penalize crimes against persons if our goal is to reduce crime by 90 percent? *How much* dark matter is there? These are questions of substantive, human significance, including the scientists in with the humans.

We Do Not Criticize Mere Amateurs

"Significance testing is difficult to do correctly. You've been looking at the attempts of amateurs or fools." No, we haven't and will not. Journal by journal, it is true, we have detected an asymmetric U-curve in the Significance Mistake. The mistake is most often committed in the less prestigious, general interest journals, whatever the field. It becomes lower, though never zero, in specialized field journals such as *Heredity, Radiology, Epidemiology, Journal of Labor Economics,* and our own *Journal of Economic History.* But then it returns to gross overuse, if not quite as much as in the low-prestige journals, in the leading general interest journals in economics, medicine, management, psychology, and the rest.

Our friend the late William Kruskal, a past-president of the American Statistical Association and coeditor of the *International Encyclopedia of Statistics,* suggested that we excavate a number of *nonpublished* articles. We agree that the exercise would illuminate the darker side of the significance-test crisis in science. But for the present it is enough to emphasize the tens of thousands of gross errors published by excellent scientists in presumably excellent journals. Our examples come purposely from the best scientists, not the worst. The journals we have quoted erect high hurdles. Economists will wait two or three years to get a rejection from the *American Economic Review.* Field by field our estimate is, so to speak, a lower bound. The crisis is at least as bad as our evidence shows. Average practice in the less mature or less prestigious regions of the sizeless sciences is sometimes, ironically, better. As Kruskal suggested, what happens to "insignificance" in the privacy of one's office is a large, dark matter needing investigation. But not here: we fixed our lower bound on the error taken from the very best journals.

OBJECTIVITY IS NOT AT STAKE

"Significance testing supplies a standard for scientific consensus. Without it, science devolves to mere opinion and crankery." That is the argument Fisher trained into his disciples, such as Hotelling and Mahalanobis and George Snedecor, and into all the inheritors of *Statistical Methods for Research Workers.* Fisher wrote:

> The value for which P = .05, or 1 in 20, is 1.96 or nearly 2; it is *convenient* to take this point as a limit in judging whether a deviation is to be considered significant or not. Deviations exceeding twice the standard deviation are *thus formally regarded as significant.* Using this criterion we should be led to follow up a false indication only once in 22 trials, even if the statistics were the only guide available. (1925a, 42; italics supplied)

And again, in Fisher's "The Arrangement of Field Experiments":

> It is *convenient* to draw the line at about the level at which we can say: "Either there is something in the treatment, or a coincidence has occurred such as does not occur more than once in twenty trials." . . . If one in twenty does not seem high enough odds, we may, if we prefer it, draw the line at one in fifty (the 2 per cent point), or one in a hundred (the 1 per cent point). Personally, the writer prefers to set a low standard of significance at the 5 per cent point, and *ignore entirely* all results which fail to reach this level. A scientific fact should be regarded as experimentally established only if a properly designed experiment rarely fails to give this level of significance. (1926b, 504; italics supplied)

And again, in *The Design of Experiments:* "It is *usual* and *convenient* for experimenters to take 5 percent as a standard level of significance, in the sense that they are prepared to ignore all results which fail to reach this standard" (Fisher 1935, 13, italics supplied). And again, in *Scientific Inference and Statistical Methods:*

> Though recognizable as a psychological condition of reluctance, or resistance to the acceptance of a proposition, the feeling induced by a test of significance has an objective basis in that the probability statement on which it is based is a fact communicable to, and verifiable by, other rational minds. The level of significance in such cases fulfils the conditions of a measure of the rational grounds for the disbelief it engenders. (Fisher 1956, 43)

Fisher, not the great transcendent, invented the 5 percent philosophy. By contrast, Gosset's economic approach to uncertainty prevented him from being able to stop thinking at .05 for fear he'd lose too much information, and profits. Fisher alone dictated the 5 percent philosophy, the Rule of Two. In doing so he turned away from Gosset and sought a mechanical, uniform, and bureaucratic line of demarcation—an "impenetrable" end, as Sartre put it, to scientific argument. So the insecure sciences, eager to establish an "objective basis" for their research "communicable to other rational minds," were pleased and materially rewarded by Fisher's 1925 transformation of the 5 percent philosophy from one of mere "convenience" into a "formal consideration." With the low fee he set for them to rise to the rank of Sciences with a big S—simply by giving institutional assent to a 5 percent philosophy—many sciences, including "consulting psychology," fancied themselves Scientific and secure.

Argument, however—not an arbitrary rule—is the heart of science. Fisher's procedure appeals to scientists uncomfortable with any sort of argument, that "indefinite approximation," as Sartre put it. To avoid debate they seek certitude such as statistical significance. The unhappy result is that mere opinion and unargued crankery are *more* likely to rule the sizeless sciences, not less. At least so it seems from the actual histories. Our own field of economics, for instance, has become in the age of computer-assisted significance testing less reasonable, not more, and it has become more, not less, swept by unexamined opinion. A technique that was supposed to end arguments has in fact merely concealed the arguments behind a facade of testing that does not test.

The folly of Fisher's Rule was demonstrated by the late Paul Meehl, a distinguished clinical psychologist and philosopher of science. Meehl showed with data taken literally from a telephone book that telephone numbers are "significantly associated" with psychometric variables. William Kruskal and Morris DeGroot pointed out repeatedly the obvious fact that "significance" can after all be attained for anything one likes merely by raising the degrees of freedom, that is, getting larger samples—a point well illustrated by the Japanese whale killers. So the folly is ancient and disturbing. Galton himself, in a seemingly serious mood, tested in 1872 the hypothesis that public prayer prolongs human life: he concluded in the negative, for why else would Queen's counsels live as long as the much-prayed-for bishops and generals?[3]

Behind Every Qualitative Question Is a Loss

Still the devout will seek a rebuttal. "Many times—and more times than you would think—the scientific problem is of a qualitative kind, and therefore the scale of oomph is not relevant." But behind every so-called qualitative question is a set of quantitative questions, each needing standards of oomph. The rhetoric of real science depends on oomph. Savage agrees.[4] Sometimes in economics (though more often in psychology, lacking the money nexus) the problem *is* a qualitative one. And then precision—a suitable amount of statistical significance, avoiding Type I error—might be part of the solution, though only a part. "Consider, for example," Savage writes, "the decision problem of a person who must buy, f_0, or refuse to buy, f_1, a lot of manufactured articles on the basis of an observation x" (1954, 252). The situation arises commonly in industrial settings, and it is not unknown in science. But even here, Savage notes, the decision to accept or reject f_0 based on observation O is really only pseudo-qualitative. Behind every buy/refuse, accept/reject, yes/no, tax/withhold, prescription/placebo, inject/don't inject, significant/insignificant qualitative judgment sits a set of quantitative questions of How Much needing answers. Does the offer f_0 minimize the maximum loss? Or does f_1? Consult the bottom line. And the bottom line in pure sciences is the persuasive oomph in the mind of other scientists.

Oomph, Not Precision, Selects the Best Model

"Significance is necessary to select among models and theories. Otherwise, how do we know which is best?" We recently had the benefit of comments on this score by some colleagues in econometrics. The occasion was a session at the January, 2004 meetings of the American Economic Association and allied social sciences in San Diego at which we made our case to a large audience of economists. Graham Elliott and Clive Granger were typical of the Model-Selection Defense at the session. They write, "[T]he original motivating experiment differentiating Einstein's theories from Newton's was the slight bending of light in an eclipse. . . . Surely not a significant effect physically for anything the theories were being used for at the time. . . . It is in this sense that . . . statistically significant effects . . . still hold interest. . . . *It may well help decide between competing theories*" (2004, 549; italics supplied).

It may, and we are very willing to concede some minor role to even mindless significance testing in science. But their instance in physics, of what was supposed to be the large, Einsteinian bending of light around the sun as against the Newtonian prediction of much less or no bending at all, is ill chosen for the point. The physicists making the experiment did not in fact use statistical tests, although they were available at the time and well within the mathematical capabilities of the physicists involved. The leader of the 1919 expeditions to photograph the eclipsed sun off the coasts of West Africa and northern Brazil to see the bending light was Arthur Eddington, the Cambridge astronomer and popularizer of relativity. Eddington, it turns out, had been a teacher of the great scientist and polymath Harold Jeffreys, who was intensely interested in the results of the expedition. (Arnold Zellner, one of several past-presidents of the American Statistical Association who participated in the 2004 symposium, has tried for decades to persuade economists and other scientists to read Jeffreys. That is partly because Jeffreys believed—against his teacher

Eddington—that statements of "existence" are for purposes of hypothesis and model testing useless.)[5]

In the Einstein-Newton debate what mattered was size, not existence. The photographic evidence was not at first persuasive. It is well known among historians of science that it took some years before an error caused by the instrumentation was corrected. Einstein's theory was at first rejected by the evidence. Eddington reasoned in favor of Einstein on geometric, a priori grounds. Jeffreys and his collaborator, Dorothy Wrinch, responded at the time with an empirically grounded criticism of Eddington's paper, published in a now famous issue of *Nature,* considering all the evidence available.[6] Size, instrumentation, design of sample, varied observations, coherence with other stories, and other sorts of argument are what decided between the competing theories. No test of statistical significance could alone shed light on the hypotheses, and didn't.

Suppose you were comparing two pieces of silverware, a spoon and a fork. Suppose you wanted to know how similar the spoon was to the fork. The procedures we and the numerous other critics in other fields are complaining about are mechanical "tests" *on the half inch of pattern on the handles, say, of each piece.* The comparison of models is reduced to the comparison of fit in the so-called sample on offer in one narrow dimension at fixed probabilities of a Type I error. Such a comparison may yield an "insignificant" difference between the spoon and the fork—imagine they come from the same silverware pattern and so have much the same figuration on their handles. But a fork on its forked end is for human uses quite different pragmatically and relative to human uses from a spoon on its spooned end. You can't stab meat with a spoon or eat soup with a fork. Precision usually does not pick the right dimension for comparison. Oomph in use does.

The Reporting Convention Is Not the Main Point

Elliott and Granger—and the rest of the commentators assembled in San Diego in 2004—agreed with our main point. Observe: a Nobel laureate in economics (Granger) agrees with our point, as did Kenneth Arrow, another laureate, on a similar occasion, in Chicago some years earlier. Yet Elliott and Granger wanted for some reason to characterize our point—which we repeat is not "ours" but that of Gosset, Savage, Kruskal, and many others—as "literary" and not "deep." They seem to believe that if the actual coefficients measuring the effects are somewhere provided in an article then "the economic [or biological or educational] significance can be determined."

Set aside that in many cases, as they admit, the articles do not provide the data to get beyond a statement that a certain coefficient is or is not "significant." Our main point is not this stylistic one. It is that real significance itself is something that needs to be argued out in the context of the scientific or policy issue at stake and cannot be determined on statistical grounds alone and certainly not (Elliott and Granger would agree) on the basis of null-hypothesis significance testing alone. Real significance cannot be "determined" by calculation, though calculation is, of course, involved in the determination. The calculation is one of many inputs—discussions of instrumentation, considerations of mathematical beauty, appeals to the authority of an Einstein or an Eddington—into what has to be a serious scientific judgment of how big is big arrived at in public.

Our point is not about matters of style, literary conventions, or superficialities of presentation. The economic significance *cannot* "be determined" by better reporting of conventional statistical tests. Elliott and Granger claim that what would be at issue in cases of bad reporting is the "statistical comprehension skills" of the reader. No, that is mistaken. It is the *economic* comprehension skills of the writer and her scientific community that matter for economic science, the *medical* comprehension for medical science, and so on. That is our main point. We cannot hand science over to a table of Student's *t,* as Gosset himself did not. Elliott and Granger join a long line of statisticians, theoretical and applied, who agree. But they hesitate.

EXISTENCE IS SELDOM IF EVER AT ISSUE

We were surprised that Joel Horowitz, who we know also agrees with much of what we say, and teaches it, asserts that "there are circumstances in which the existence of a phenomenon, not its magnitude, is decisive" (2004, 551). Horowitz, unlike us, was trained as a physicist. But here he is talking like a mathematician. We ourselves favor the talk of physicists such as Richard Feynman, whose great "elementary" textbook at Cal Tech is filled with statements such as some magnitudes "are zero, *or can be neglected in comparison with the variations in the other directions*" (2:7–2, italics supplied) or "[T]he fact that there is an amplitude . . . has *little effect* when the two positions have *very different* energies" (3:9–8, italics supplied). Or in his lectures on computation that "Predictive coding enables us to compress messages to *a quite remarkable degree"* (1984–86 [1996], 129).

Horowitz will be able to tell us what Feynman was talking about as far as the physics is concerned. But what is obvious in Feynman's talk even to an outsider is that it is always about magnitudes, never about existence in the mathematician's sense. A mathematician trying to prove that a number is greater than zero doesn't give a fig whether the number is 10^{100} or 10^{-100}. Zero is zero, and anything greater than it is . . . greater than zero. The physicist, however, does care about how big the number is, every time . . . all right, we concede our lack of expertise in physics and will grant Horowitz his rare example: *nearly* every time. Think of the famous moment at which Feynman tested with a glass of cold water, a clamp, and O-ring material a hypothesis about the explosion of the space shuttle *Challenger.* Was a temperature *around* freezing, Feynman asked and answered with devastating cogency, *low enough* to change the behavior of the material used for the O-rings—*low enough to matter?* Yes. The O-ring stiffened then broke. And, although Feynman's sample size was merely $N = 1$, every scientist considered the case closed.

Horowitz gives a physical example by Cronin and Fitch of mere existence mattering. But presumably if the effect Cronin and Fitch sought had turned out to be two orders of magnitude greater than it was in fact, then the surrounding physics would have been greatly altered. So magnitude mattered even in their case of a very faint effect. And, as Horowitz himself notes—exhibiting, incidentally, a physicist's sensitivity to error bounds—economics is commonly not precise enough for tiny effects to be relevant anyway, a point made half a century ago by Oskar Morgenstern in his *On the Accuracy of Economic Observations.*

The problem with Horowitz's existence talk—which, we repeat, we do not think even he believes is very important in most scientific practice, since on the whole he agrees with us and raises our point with his students—is that it suggests there must be a precision-oriented "test" for it free of any worries about how big is big. But there isn't. An alternative hypothesis, somewhere near to or far from the null, ensures it. Horowitz says that "the difference [found in a study of intergenerational wage transmission] between [0.2 and 0.4] . . . is interesting and important only if we can be reasonably sure that it is not an artifact of random sampling error" (Horowitz 2004, 552). He is applying against his better judgment an arbitrary criterion of statistical significance—which after all is the main thing both he and we don't like. The point is that even if (say) a 95 percent confidence interval contains both 0.2 and 0.4 that doesn't mean there "exists no difference" between the two numbers or that we are justified in thinking there is "no difference" in the predictions of the 0.2 theory and the 0.4 theory. It depends on the loss function every time. To put it another way, it depends on the significance level one chooses relative to alternative hypotheses, and even that (as Gosset and Neyman and Pearson stressed) is a scientific and social decision not to be left to meaningless ritual disguised as a convenience or a merely formal convention.[7]

If you were required to make from the 1992 wage study some crucial decision, and had estimated a coefficient of 0.4, though alas, from data noisy when considered as a random sample, you might have to go ahead and suppose that 0.4 was the Truth. (In fact the coefficient of 0.4 in question, from Gary Solon's exemplary 1992 study of intergenerational income mobility, passed conventional tests of statistical significance and seems, moreover, to be the product of a Pareto improved model for extracting income parameters.)

We agree instead with the point made at the session by Edward Leamer (2004) that science needs tests of persuasiveness or usefulness, both of which could be called in official philosophical language "pragmatism." "Models are neither true nor false," Leamer, a pioneering econometrician, writes. "They are sometimes useful and sometimes misleading. *The goal of an empirical economist,*" he continues, "*should not be to determine the truthfulness of a model but rather the domain of its usefulness. . . .* Does the parable of the Invisible Hand persuade?" he asks rhetorically. "How does that compare with the parable of the *t*-value? I think the first is compelling but the second is utterly fanciful and completely unpersuasive"[8] Yes.

And Leamer is also right when he points out that tests of significance persist precisely because, though precise, they do not in fact settle much in the matter of usefulness. Consider the enormous number of tests of significance performed each year on both sides of every issue in economics. Our mailboxes are overflowing with fat articles on the "significance" of strategic tariffs and the "insignificance" of welfare subsidies. Would it surprise anyone to assert that the tests were annually, let us say, on the order of one hundred thousand, a mere one thousand economists performing a hundred tests each? But some economists we know perform more than a hundred tests before lunch. Worldwide a better guess is—and here we include economists employed in a variety of business, government, and university occupations—ten million. Ten million tests of significance, in economics, annually. If the ten million tests were in fact as conclusive as their own rhetoric requires, whether accepting or rejecting, then nearly every issue in economics would long since have been settled. By now there would therefore be far fewer tests per year, not, as is the case, more and more.

In any event we are confident that both Leamer and Horowitz would agree with us that when one wants to compare a spoon and a fork it would be wise to develop other ways of comparing them than statistical tests on the design of the handles, over and over and over again, ten million times.

JOURNAL SPACE IS NOT THE PROBLEM

"There's simply no room on the published page for interpretation of substantive significance. The job of the author is to show the 'significant' facts and let the reader decide the rest." But if in a study of salmonella in Greek swineherds or of price inflation and the federal funds rate or of Valium and clinical depression you have chosen to isolate variables and models on the basis of their statistical and not their substantive significance, the reader is not equipped to judge. Inability to judge is rigged by a concealing of the objects of interest.

And the reader is not in any case equipped for judgment by the statistics provided. Every year many thousands of published articles, we have noted, do not supply the units of measurement or the table of descriptive statistics necessary for the computation and interpretation of effect size, "statistically significant" or not. Journal space is scarce, to be sure. But allocating table and paragraph after table and paragraph to *statistical* significance, with no context of the units or the means involved, is not efficient. Savage saw the fundamental problem in 1954: "Though modern objectivistic statisticians may recognize the existence of differences of judgment, they argue in theoretical discussions that statistics must be pursued without reference to the existence of those differences, indeed without reference to judgment at all" (1954, 156).

HOW SCIENTISTS SOLVE THE PROBLEM OF *Sorites*

"But I don't know what the *magnitudes* you keep talking about should be. How can I judge the line between large and small? Statistical significance gives me a way to decide that doesn't involve judgment." Such a reply is grievously misled. Yet it is the most common reply, even among our cost-conscious and numbers-rich colleagues in economics.

A judgment must be made somewhere. If you rely on statistical significance you are making a judgment without judging. The procedure merely hides the judgment, even from yourself. You might as well use a table of random numbers or a late-night phone call to a foreign-language psychic network to make your scientific decisions.

Real scientists draw a line between large and small. In science one always faces the ancient problem of *sorites* (sore-*it*-eez), a Greek word meaning "heaped up."[9] What is the maximum height of a "short" person? What is the maximum amount of hair on a "bald" man's head? What is the smallest half of a "half" cookie? How long is a "long" life? Max Black, a well-known American philosopher, wrote, "A man whose height is four feet is short; adding one tenth of an inch to a short man's height leaves him short; therefore, a man whose height is four feet and one tenth of an inch is short" (1970, 1). By repeatedly "heaping on" additional tenths of an inch you reach the paradox. Thus a man whose height is four feet and *two* tenths of an inch is short and so forth. Heaping repeatedly you eventually ask, "Where can I draw the line? At what point does a man stop being short and start being something else, such as medium or tall? And what does it matter?"

Scientists make a judgment about a threshold and get on with it. They are not afraid to establish thresholds of oomph and then get on with the task of estimating and interpreting the mattering question. So to the immediate question "What height makes for a tall statistician?" a preliminary oomphful response is: "William Kruskal was short. When a statistician attains the height of Stephen Stigler, then he is definitely tall." But the answer you give to the argument from *sorites* depends of course on the special purposes of your investigation. "Short" or "tall" for what purpose? For purposes of sitting comfortably in a standard library chair? For purposes of taking dishes down from the top shelf of a standard American cupboard? For purposes of marital matching with Dutch people in 2007? For purposes of comparing the heights of Mesoamericans before and after the introduction of a corn-based diet?

It is the kind of question the immunologist and Nobel laureate Peter Medawar had to answer when he examined *how much* transplanted tissue the human body could tolerate without producing prohibitively bad side effects—a quantitative standard. It is the kind of question the economist A. B. Atkinson had to answer when he examined the minimum standard of living the average person would need to actively participate in civil society—a quantitative standard. Fisher's claim that there is some extrascientific way of deciding oomph in the probability numbers themselves, without regard to human purposes, is mistaken.

So Bright under the Lamppost

"I'm just trying to see if there's anything *to* the theory. Later on I'll test its economic significance." But a *t*-test can't do the first step. And "later," we repeat, usually never comes. A man walking at night comes upon a friend, drunk, crawling around under a lamppost. "I'm looking for my keys," says the drunk. "Oh, all right, let me help you," says the friend, and asks, "You lost them here under the lamppost, yes?" "No," replies the drunk, "I lost them out there, in the dark. . . but the light is so much brighter under the lamppost." The sizeless scientists construe every question of real significance as being a question of Type I error concerning a random sample because the light of the standard error is so wonderfully bright. The scientific objects therefore lie there still, undetected, in the dark.

To which at last the sizeless scientist exclaims in anguish: "Precisely! The dark is so very dark. What are we to *do* if not calculations of statistical significance?" We reply: you are to *do* real science. Real science, unlike significance-testing science, is difficult. If it were not, it would not be real science, but instead it would be already established routine. Real science asks you to make real scientific judgments and real scientific arguments within a community of other scientists. It asks you to be quantitatively persuasive, not to be irrelevantly mechanical. Life is hard.

But we understand your generous impulse to doubt that Ziliak and Mc-Closkey can be justified in their indignation about the meaninglessness of statistical "significance." Even though you acknowledge the badness of its practice, and can't see what is wrong with our numerous arguments, you are still uneasy.

Adam Smith, who was much more than an economist, noted in 1759 that hatred, resentment, and indignation against bad behavior serve, of course, a social purpose, for "the utility of those passions . . . to the public, as the guardians of justice, [is] . . . considerable" (35). "Yet there is still something disagreeable in the passions themselves, which makes the appearance of them in other men the natural object of our aversion." He explained that the impression on the impartial spectator of the indignation is present and vivid, but the impression of the social gain from the curtailing of the bad behavior is remote and feeble. "What our sympathy for the person who feels [the indignation against statistical significance] would prompt us to wish for, our fellow feeling with the other [namely, the sizeless scientist] would lead us to fear. As they are both men, we are concerned for both" (34). Since "it is the immediate and not the remote effects of objects which render them agreeable or disagreeable to the imagination," you feel sympathy for the probably very nice, if not harmless, Sizeless Scientist (35).

Wise and just Adam Smith. We honor your sympathy for the subjects of our indignation. Nevertheless we want you to adopt the indignation. We realize it is emotionally difficult. But for your own good we urge you to try. Try to bring to your imagination the not so remote badness of sizeless science and the social gain from curtailing it.

4

Better Practice:
β-Importance vs. α-"Significance"

> We have the right to believe at our own risk any hypothesis that is live enough to tempt our will.
>
> WILLIAM JAMES 1896, 107

> I have no Faith in anything short of actual Measurement and the Rule of Three.
>
> CHARLES DARWIN, N.D., FRONTISPIECE,
> INAUGURAL ISSUE, *Annals of Eugenics* (1925)

> Yet in withholdings of specification I could but betray you worse still.
>
> JAMES AGEE AND WALKER EVANS 1941, 101

We are not by any means the first people, even in economics, to express such indignation (Edgeworth 1885, 215; Morgenstern 1950, 92–93; Savage 1954, 1971a; Arrow 1959; Tullock 1959; De Finetti 1971; Leamer 1978; Mayer 1979; Zellner 1984).

And in most fields there have been others. In *psychology and education:* Boring 1919; Tyler 1931; Sterling 1959; Cohen 1962, 1994; Edwards, Lindman, and Savage 1963; Meehl 1954, 1967, 1978, 1990, 1998; Bakan 1966; Rozeboom 1960, 1997; Walster and Cleary 1970; Shulman 1970; Carver 1978, 1993; Gigerenzer et al. 1989; Thompson 1996, 1997, 2002a, 2002b, 2004; Schmidt 1996; Robinson and Levin 1997; Vacha-Haase et al. 2000; Hubbard and Ryan 2000; Fidler 2002; Fidler et al. 2004a, 2004b; Altman 2004. And *business and operations research:* Student 1904, 1905, in E. Pearson 1939; Deming 1938, 1961, 1982; Neyman 1961; Raiffa and Schlaifer 1961; Polley 2003. And in many other fields, such as *law:* Kaye

2002; McBurney and Parsons 2002; Steward and O'Donnell 2005. *Sociology and political science:* Woofter 1933; Morrison and Henkel 1969, 1970; Cain and Watts 1970. *Archaeology:* Rosen 1986; Orton 1997. *Marine, wildlife, and conservation biology:* Peterman 1990; Taylor and Gerrodette 2004; Csada, James, and Espie 1996; Hayes and Steidl 1997; Parkhurst 1997; Anderson, Burnham, and Thompson 2000; Robinson and Wainer 2002; McCarthy and Parris 2004; Cole and McBride 2004; Fidler et al. 2004a, 2004b. *Public health:* Rossi 1990; Matthews 1998; Fidler et al. 2004a, 2004b. *Epidemiology and medicine:* Cutler et al. 1966; Freiman et al. 1978; Goodman 1992, 1993, 1999a, 1999b; Lang, Rothman, and Cann 1998; Rothman 1978, 1986 (chaps. 7–10), 1998, 2002; Rothman, Johnson, and Sugano 1999; Rennie 1978; Savitz, Tolo, and Poole 1990; Altman 1991, 2000; Sterne and Smith 2001; Fidler et al. 2004b. *Sports medicine:* Batterham and Hopkins 2005; Marshall 2005. And in *statistics* proper: Student (W. S. Gosset) 1904, 1905, 1926, 1927, 1929, 1936 (in Pearson 1939), 1937, 1938, 1942; E. Pearson 1926, 1939; Neyman and Pearson 1928, 1933; Pearson and Adyanthaya 1929; Jeffreys 1939a; Wald 1950; Yates 1951; Yates and Mather 1963; Deming 1938 (30), 1961 (55–57), 1982 (369), Wallis and Roberts 1956; Neyman 1956, 1957; Kruskal 1968a, 1968b, 1978, 1980; Tukey 1969; Atkins and Jarrett 1979; Lehmann 1959; Good 1981, 1992; Lindley 1991; Zellner 1984, 1997, 2004a; Moore and McCabe 1993; Harlow, Mulaik, and Steiger 1997; Berger 2003; Gelman and Stern 2006. And on and on and on.

William L. Thompson, a statistical and environmental biologist with the National Park Service in Alaska and the author of numerous books on ecology, recently compiled a bibliography of "326 Articles/Books Questioning the Indiscriminate Use of Statistical Hypothesis Tests in Observational Studies"—fully 326, ranging over dozens of fields. He concludes gloomily:

> Unfortunately, this approach was (and continues to be) pounded into us at both the introductory and advanced level of statistics in universities throughout the world. The general lack of awareness of problems with statistical hypothesis testing is especially acute in my own field (ecology/environmental science/fish and wildlife biology). . . . This number of articles [namely, the 326] pales in comparison to the vast array of articles devoted to [using and teaching the technique] . . . in the social, medical, and statistical sciences. Indeed, until very recently, I was one of the "unaware" who blindly applied statistical hypothesis tests to observational data without considering the validity of such an approach. (http://www.cnr.colostate.edu/~anderson/thompson1.html)

All the serious students of the matter say with Thompson: for Lord's sake, reflect on the substantive significance. Don't put the hypothesis through a screen of arbitrary levels of t or p or R^2. Flee the fallacy of misplaced concreteness and its wicked twin, the fallacy of the transposed conditional. Eschew testimation—that unhappy marriage of the transposed conditional and the sizeless stare.

Despite hundreds and hundreds of such warnings, in none of these fields has practice changed. Not a bit. Testimation rules. Near the end of his life, in 1956, Fisher backed away from his rule, reverting to a procedure advocated long before by Karl Pearson—that researchers should fix their own levels of significance (1956, 41–46). Fisher did not go so far as to adopt loss functions or any other economic approach to inference. But anyway by 1956 no one was listening. His converts had their faith of 5 percent.

You can be cynically relaxed about the situation, and argue amiably that it doesn't matter, because after all it's merely eighty years of academic talk. "Of course, everyone knows that statistical significance proves approximately nothing," you can say with a superior smile. "But the error is mostly propagated by scholars in print in their scholarly journals. No real world decision depends on it." A colleague of ours in economic history has said essentially that (O'Brien 2004, 565). Others say with that same smile: "It's just a game, a way of getting a publication and a promotion."

We think that such amused cynicism will not do. For example, it's not true that significance is "just a game." Real world decisions *do* depend on Fisher-circumscribed significance—witness the Vioxx debacle, the orphaned lambs, the slaughter of minke and sperm whales in Antarctica, the U.S. appellate courts, and thousands of other practical decisions large and small.

Be "The Humble Applier of Formulae"

Gosset wrote in 1929 a characteristically graceful letter admonishing his difficult friend R. A. Fisher.

> I think you must for the moment consent to be analyzed into α-Fisher the eminent mathematician and β-Fisher the humble applier of his formulae. Now it's α-Fisher's business, or I think it is, to supply the world's needs in statistical formulae: true, . . . α-Fisher is interested in the theoretical side and β-Fisher in whatever seems good to him. But when β-Fisher says that the detailed examination of the data is his business and proceeds to examine them by means of tables which are strictly true only for

normally distributed variables I think I'm entitled to ask him what difference it makes if in fact the samples are not taken from this class of variables. (Gosset to Fisher, Letter no. 104, in *Letters of William Sealy Gosset to R. A. Fisher, Vols. 1–5*, 97–98)

Be a humble applier of formulae, Gosset says. But then judge. Ask what difference the effect size makes. Pay homage to your β-self.

The α-scientist is concerned with the promise of asymptotic results, the holiness of the normal law, the precision of the *t*-distribution, the scale of Type I error. The β-scientist is, by contrast, concerned with empirical interpretation and judgment, "whatever seems good" to her and the scientists she is in conversation with. She is concerned with the size and substantive significance of the regression coefficient, with the small-sample experience of life, and with the prevalence of nonnormal distributions. She is concerned with prior probability and Type II error—the power of the test—but understands that priors and power do not alone entail substantive significance, always a matter of scientific or policy judgment, "what difference it makes." Power, as the epidemiologist Kenneth Rothman has warned, can itself be used as a backdoor way to substitute statistical significance for size-matters/how-much and therefore should be used gingerly (Rothman 1986, 79–82). Gosset would have agreed.

But the β-approach takes humility. Fisher was, Gosset believed, aware of his β-self. Fisher did not employ it with a fraction of the energy he devoted to his α-self. "The good man," wrote the novelist and philosopher Iris Murdoch, "is humble; he is very unlike the neo-Kantian Lucifer. . . . Only rarely does one meet somebody in whom [humility] positively shines, in whom one apprehends with amazement the absence of the anxious avaricious tentacles of the self" (1967, 103). Murdoch points out that humility is one of the chief virtues in a good artist and a good scientist. In his *Justice as Translation* the legal scholar James Boyd White put it in terms of humble reading, "a willingness to learn the other's language and to undergo the changes we know that will entail" (1989, 42).

A complete scientist will keep his α- and β-selves in analytical and ethical balance. James Agee, a great American poet, saw the reason, writing that "in withholdings of specification I could but betray you worse still" (Agee and Evans 1941, 101). In the twentieth century the α-only specifications have betrayed science. Specification problems, as β-Leamer (1978) and others have noted, will always exist. But Gosset and the poet were plain: it is impossible to "solve" even the errors of specification by ignoring the substantive side, the β of coefficients and Type II error.

WITHOUT β, SCIENCE IS BLIND

"I chanced on a wonderful book by Marius von Senden," wrote the essayist Annie Dillard, "called *Space and Sight.*"

> When Western surgeons discovered how to perform safe cataract operations, they ranged across Europe and America operating on dozens of men and women of all ages who had been blinded by cataracts since birth. Von Senden collected accounts of such cases; the histories are fascinating. Many doctors had tested their patients' sense perceptions and ideas of space both before and after the operations. *The vast majority of patients, of both sexes and all ages, had,* in von Senden's opinion, *no idea of space whatsoever. Form, distance, and size were so many meaningless syllables.* A patient had no idea of depth, "confusing it with roundness." Before the operation a doctor would give a blind patient a cube and a sphere; the patient would tongue it or feel it with his hands, and name it correctly. After the operation the doctor would show the same objects to the patient without letting him touch them; now he had no clue whatsoever what he was seeing. One patient called lemonade "square" because it pricked on his tongue as a square shape pricked on the touch of his hands. (1974, 25; italics supplied)

The patient blinded by cataracts is forced to live in a world without size or the perception of it. But the sizeless scientist *chooses* to focus on precision rather than oomph and to call lemonade "square."

But, really, this inability to see size can't last. As Dillard put it, surrounded as she was by valleys and hills and orchards:

> I couldn't sustain the illusion of flatness [the null]. I've been around for too long. Form is condemned to an eternal danse macabre with meaning: I couldn't unpeach the peaches. . . . I live now in a world of shadows that shape and distance color, a world where space makes a kind of terrible sense. What gnosticism is this, and what physics? (1974, 29–30)

Sizeless scientists "sustain the illusion of flatness"—whether accepting or rejecting the null hypothesis—at the expense of science. It can't go on.

5

A Lot Can Go Wrong
in the Use of
Significance Tests in Economics

There remains no consensus on whether to use P, p, P, *or* p *values.*

D. G. ALTMAN 1991, 1902

THE MISEDUCATION OF YOUR AUTHORS

We are quantitative economists. Our scientific work has concerned British industrial history and medieval agriculture, labor market statistics and the history of American charity, all viewed quantitatively. It has asked, How Much?

We grew up, though, with statistical significance. McCloskey was a statistical student of Guy Orcutt and John R. Meyer at Harvard in the 1960s. For her oral exams she presented herself, laughably, as an econometrician. She should have failed, considering her inability in 1966 to answer an elementary question about . . . statistical significance. By the ramped-up standards thirty years later, Ziliak, a student at the University of Iowa, was well trained econometrically. But he, too, committed the standard error, making industrial employment forecasts in the late 1980s with the aid of the broken instrument. Taken together we believed in the mistaken procedures of statistical significance for about twenty-two years, a third of our combined careers, nearly a quarter century of bald testimating.

It could have been longer if after a dozen years or so as an economist McCloskey had not heard a hint in the teachings of Richard Zecher or if near the beginning of his academic career Ziliak had not gotten that phone call about black unemployment rates, or seen a glimmer of hope on a page of Wonnacott and Wonnacott 1982 (160) and then a brightly blinking warning light in *The Rhetoric of Economics* (McCloskey 1985b). Since realizing that the procedures are flatly mistaken we have said so in print, McCloskey starting in the early 1980s, joined by Ziliak in the early 1990s, and anticipated by the handful of statistically savvy economists we have mentioned, Arrow to Zellner.

By the 1990s the very few sizeless economists who reacted at all to our doubts replied in effect, "Yes, I know it's silly to think that fit is the same thing as substantive importance. But *I* myself don't do it. Only those numerous bad economists do." We knew this to be false. Any professional economist self-conscious about the rhetoric of the field knows that statistical significance rules with bad economists—whoever those are—and hard.

Significance rules, admittedly, in a strange way. Economists do not appear to actually *believe* the results certified as "significant." They believe the results of their science on other grounds: symmetry, coherence, parsimony, motivation, novelty, oomph. These and other virtues in action change minds. No one's mind appears to change when a scientific opponent offers a statistically significant finding that, say, the minimum wage increases rather than decreases employment, one of millions of such tests performed yearly. When an economist appears with a table showing, let's say, a "significant relation" between Catholicism and economic backwardness, some will cheer, some will boo. But no minds will be changed. The offering of statistically significant coefficients seems ceremonial.

Yet it is a ceremony that economists do not want to give up. All the leading econometricians we have encountered, of course, agree with our point in substance. After all, the point is trivially, obviously true. Whatever the value of a consideration of Type I error and measures of fit in a properly drawn random sample may be for this or that supplementary and auxiliary purposes, a low Type I error and a high statistical fit are obviously not *the same thing* as scientific importance. A merely statistical significance, obviously, cannot substitute for the judgment by a scientist and her community about the largeness or smallness of a coefficient, judged against standards of scientific or policy oomph. As Harold Jeffreys told his fellow physicists long ago, to reject a hypothesis because the data show "large" departures from the prediction "requires a quantitative criterion of what is to be considered a large departure."[1] Yes. Scientific judgment requires quantitative *judgment,* we say yet again, not endlessly more complicated machinery. The machinery is very handsome, and sometimes even useful. But at the end of the day, having skillfully used it, the economic scientist needs to *judge* its output.

The economists and calculators of the 1990s replied to us, "Don't fret. It's only those *stupid* economists who make the mistake," the mistake of substituting mechanical tests for judgment. We realized, with some irritation considering the effort involved, that the only way to respond to such a reply was to measure. As quantitative scientists that was our inclination anyway.

Meet Dr. Neyman Pearson

In 1996, therefore, we published in a review journal in economics an article called "The Standard Error of Regressions." The title was another of our attempts to be witty, since what we were complaining about was precisely the reduction of scientific judgment to a rule of standard error (i.e., the standard deviation of the estimate of a coefficient in a regression). But in the article we did more than complain theoretically because a dozen years of our theoretical complaints had had by 1996 no discernible effect at all on econometric practice.

Even in our own subfield of economic history, where McCloskey and then Ziliak made the significance point early and often, only a very few grasped it firmly enough to change their practices. They had a vague feeling that McCloskey and Ziliak "don't like the word *significance.*" But that was all. McCloskey remembers, for example, a leading economic historian *whispering* repeatedly in McCloskey's presence the very words "statistical significance." The eminent historian was giving a talk at a little conference they both attended in Italy in the early 1990s. The sarcastic whispering was meant to show that he realized she didn't like his rhetoric of significance but he did not understand why. Further, he wanted it to be clear that he did not think her silly dislike was worth trying to understand. But of course it's not about the word. Despite many attempts, public and private, to explain the point to him, and despite the man's scientific excellence on many other matters, he appears never to have grasped it.

In this the historian is typical of numerous other colleagues we admire as economic scientists. At a conference on economic history in Kansas in 1995 Ziliak, who was at the time a graduate student, complained similarly about an otherwise fine paper by a former student of the whispering historian. Like his teacher the student relied exclusively on the sizeless stare.

One of the older pioneers of historical economics was the chair of the session. The chair cut Ziliak off at mid-sentence, saying to the audience of some fifty economic historians, "Let's get back to the important questions." *Laughter* was the result, followed by more sizeless staring. The older man is a great economic historian, too. But, like the whispering one, and his own former student, he has never grasped the point about the unimportance of merely statistical significance. Some of his otherwise astonishingly fine scientific work has been undermined by failing to appreciate it.

After the session, Ziliak asked the author of the paper if he had considered framing his statistical problem as a Neyman-Pearson decision. The young professor replied, "No, I haven't. But who's Neyman Pearson?"

In the face of evasion and plain ignorance we decided to measure the practice of the best economists, defining "the best" as "those who published full-length articles using regression analysis in the *American Economic Review* during the decade of the 1980s." We admit that this "best" talk is a bit foolish. Many of the best economists do not write for the *AER*. But every economist knows what an article in the *AER* represents in terms of career by the mechanical standards by which academic quality is judged these days, a standard of best persuasive to deans. It is a corrupt and silly standard and has relieved economists of actually having to *read* anything by the candidate for a hire or promotion (McCloskey 2002). Such is the magical power of deans to corrupt science. Still, economists show by such behavior that they think a publication in the *AER* is the gold standard.

The 182 applied econometric articles published in the *AER* in the 1980s did not constitute a random sample of anything. They were the entire universe of such articles or else they were a nonrandom sample of elite articles of a certain middling technical level in econometrics. So an argument from sampling, that wonderful ratio $\sigma/\sqrt{n}$, was not relevant to the measurement. We do not here use significance testing.

During the 1980s the *AER* also published shorter articles printed in smaller type. We decided not to include these articles because they were somewhat miscellaneous, some being mere comments on the full-length articles. The decision was fortunate, since in the *AER* of 1984 McCloskey coauthored just such an article, a (wonderful) exploration of storage costs and interest rates for grain in medieval England and in it grossly and stupidly misused . . . statistical significance.

THE QUESTIONNAIRE

We asked nineteen yes/no overlapping questions about statistical practice in each of the 182 articles. The questions are posed so that a yes answer means that the practice is good, "good" by the standards of any serious statistician. The criteria are not controversial. We are not measuring here the other problems with statistical significance, only the noncontroversial and elementary errors in its handling that are the subject here. We do not measure, for example, the selection bias in articles for publication on the basis of the significance levels, the application of sampling theory to non-samples such as entire populations, or, more controversially, the application of sampling theory to entire instantiations of a time series. We do not consider all the sophisticated problems of testing and estimation that numerous statistical scientists have worried over, Gosset to Manski. Suppose for the moment that these other problems have been solved or at any rate set aside. As the humanists say, we "bracket" the other, more advanced problems.

Even people who are not yet persuaded that null-hypothesis significance testing has ruined statistical economics, in other words, will agree with the standard implied in our questionnaire. No competent statistician would recommend, for example, that economists use *only* tests of statistical significance without a loss function or a consideration of power or that they *not* report the quantitative scale along which an estimated coefficient is to be considered large or small. We were encouraged to find that Fidler et al. (2004b) employed a similar questionnaire for their study of best practice in medicine and psychology, unaware of McCloskey and Ziliak 1996.

So the percentage of articles that answer yes is a pretty good measure of the elementary goodness of econometric practice in the matter of statistical significance in the leading general-interest journal of economics. We do not claim it is perfect—for instance, we regret that we did not ask about confidence intervals or specification error or Bayes's rule—or that our application of it is perfect. We have tried to be fair, but another economist might code this or that article a little differently. We don't think it will matter a great deal to our findings, but we welcome alternative measures and recoding of the articles. The articles are there in public view, and we invite people such as Hoover and Siegler (2008) who doubt that Fisherian practices dominate economics to try their hand at replication.

The nineteen questions are as follows.

1. *Does the article depend on a small number of observations such that statistically "significant" differences are* not *forced by the large number of observations?* Little is learned from a finding that a certain regression coefficient is "significantly different from zero" if the sample size—corrected for degrees of freedom—is two hundred thousand. With very large samples, *every* variable passes *any* given test of having a low standard error of the estimate of the mean, *s*, at any rate if one makes no correction for sample size. Against the better judgment of the William Gossets and Morris DeGroots, if the coefficient on a variable is insignificant, "no worries" seems to be today's prepublication attitude: merely increase the *N* to get a still lower *s*. It is therefore pointless in such circumstances to bring in statistical significance as a "decisive" factor. You know before looking at the data that every coefficient will be statistically significantly different from the null. Notice the implication of such reasoning. It implies that something must be very wrong with the notion that statistical significance is *necessary* for substantive significance, a preliminary screen through which one puts one's data. For example, in the 1980s article by Blomquist, Berger, and Hoehn their *N* was 34,414 housing units and 46,004 people (*AER,* March 1988, 93). With such sample sizes a variable that is economically unimportant will show up as statistically significant, through the sheer force of large *N*. Blomquist, Berger, and Hoehn used Fisher's Rule of Two despite their gigantic samples. So the article, a fine one in many other ways, earns on this question a No.

2. *Are the units and the descriptive statistics for all regression variables included?* The slope of an equation in which the units of the variables are not included cannot be interpreted (as Elliott and Granger [2004] note and as is anyway obvious). No one can exercise judgment as to whether something is importantly large or small when it is reported without units or a scale along which to judge them large or small at "−6.8" or "567.987." Inches or pounds? Logs or levels?

3. *Are the coefficients reported in elasticity form, or in some interpretable form relevant for the problem at hand, so that the reader can discern the economic impact?* In their theoretical work economists commonly avoid the problem of units by speaking of elasticities, that is, percentage changes in one variable with respect to percentage changes in another. This is better than leaving the units in natural units, not supplied (as in question 2). But often an article will not give the actual magnitude of the elasticity but merely state with satisfaction its statistical significance.

The statisticians Wallis and Roberts noted long ago that "sometimes authors are so intrigued by tests of significance that they even fail to state the actual *amount* of the effect, much less to appraise its practical importance" (1956, 409; italics in original). Since 1956 the error has become a good deal more prevalent than Wallis and Roberts's choice of the word *sometimes* would suggest. To appraise an effect you need a quantitative standard of oomph, and to appraise the standard of oomph you need interpretable coefficients.

4. *Are the proper null hypotheses specified*? Sometimes the economists will test a null of zero when the economic question entails a null quite different from zero, of 1.0, say, or $\beta_1 + \beta_2 = 0$. The economic question may be about a unit income elasticity of demand for money, for example, but the article will report "significance" as difference from *zero*. A canned program, testing automatically against zero, is making the scientific decision. In a 1990 Sage pamphlet the political scientist Lawrence B. Mohr recommended the mechanical test against zero as follows: "[W]e just want to know whether or not a certain relationship . . . is worth further thought. . . . If the relationship. . . is zero or close to it, then it does not merit further exploration"(1990, 8). He is mistaken. For one thing, the "further thought" has to take place relative to the appropriate null and its relevant alternatives, which are not always and everywhere zero. For another, an effect "close" to zero on some scale, rejected at the 5 percent level of significance, can be substantively important. But it depends on the loss function every time. The numbers are close to zero, you claim to an options analyst trading with millions. How close is close?

5. *Are the coefficients carefully interpreted*? The econometrician Arthur Goldberger explains the point by imagining that the dependent variable in a regression is weight in pounds and the independent variables are height in inches and amount of exercise in, say, miles walked per week (1991, 241). Suppose the coefficient on height is statistically significant at $p < .05$ while the coefficient on exercise is not, though large in magnitude and negative in sign. A doctor would not say to a patient, "The problem is not that you're fat—it's that you're too short for your weight." The exercise variable is the policy-relevant one (as may be other, omitted variables such as dietary habits). A test of significance on height is in such a case beside the point.

6. *Does the article refrain from reporting* t- *or* F-*statistics or standard errors even when a test of significance is not relevant*? A No on this ques-

tion is another sign of canned regression packages taking over the mind of the scientist. A significance test on an entire population is an example of such alien invasion.

7. *Is statistical significance at its first use merely one of multiple criteria of "importance" in sight?* Often the first use will be at the crescendo of the article, the place where the author appears to think she is making the crucial factual argument. But statistical significance does not imply substantive significance. In other words, something is wrong with the notion that statistical significance is *sufficient* for substantive significance. Note that this is a different criticism than saying that something also is wrong with the notion that it is *necessary.* Both criticisms hold: statistical significance is neither necessary nor sufficient for substantive significance. The two criticisms add up to a simple statement of our point: statistically "significant" variables can be substantively speaking unimportant and statistically "insignificant" ones can be substantively speaking important. Articles were coded Yes if statistical significance played a second or lower-order role, at any rate below the primary considerations of substantive significance.

8. *Does the article mention the power of the test?* For example, Frederic S. Mishkin does, unusually for an economist in the 1980s, though he does so in footnotes.[2] In capital market studies a lack of power is a common problem seldom faced because seldom mentioned.[3]

9. *If the article mentions power, does it do anything about it?* Mishkin does not. The first empirical estimation of a power function was carried out by Egon Pearson and N. K. Adyanthaya, in a 1929 issue of *Biometrika*. In the 1950s Jerzy Neyman and Elizabeth Scott estimated power functions in their articles on galaxy clustering and weather patterns. But Fisher's studied ignorance of Type II error (as late as 1955 and 1956) won over Harold Hotelling, most of Hotelling's students, and most of the rest of the economics profession. Exceptions include Wald, Savage, Zellner, Horowitz, Eugene Savin, Allan Würtz, and Robert Shiller. Among the very few pieces of economics that theorize or estimate power functions are Zellner 1984; Shiller 1987; Horowitz, Markatou, and Lenth 1995; Savin and Würtz 1999; Horowitz and Spokoiny 2001; and Würtz 2003. It is our impression—perhaps someone can offer an alternative hypothesis—that most economists avoid considerations of power because they don't understand it. But as Neyman (1956, 290) remarked long ago, without a measure of power "no purely probabilistic theory of tests is possible."

10. *Does the article refrain from "asterisk econometrics," that is, ranking the coefficients according to the absolute size of their t-statistics?* In an article published in the next decade of the *AER,* the current Federal Reserve Board chair Ben Bernanke and former Council of Economic Advisers and FRB member Alan Blinder do not (Bernanke and Blinder 1992, 905, 909). Ranking coefficients by the absolute size of *t* and regardless of coefficient effect is especially common in psychology and finance.

11. *Does the article refrain from "sign econometrics," that is, noting the sign but not the size of coefficients?* The distribution-free "sign test" for matched pairs is on occasion scientifically meaningful. Ordinarily sign alone is not *economically* significant, however, unless the magnitude attached to the sign is large or small enough to matter. It is not true, as "sign econometrics" supposes, that in parametric studies sign is a statistic independent of magnitude. Sign econometrics originated in economics with Paul Samuelson's mathematics and philosophy department claim in 1947 that economics could be a science of *qualitative* predictions. Samuelson might have followed instead the oomph-talking departments of physics or engineering he migrated from. More precisely, he could have adopted Gosset's "real error bars" or at least Neyman's "confidence intervals." Edgeworth made the point about sign back in 1885 in the article that coined the very term *significance* (1885, 215).

12. *Does the article discuss the size of the coefficients at all?* Once regression results are presented, does the article ask about the *economic* significance of the results? In other words, is there a discussion of the policy or scientific importance of the fitted coefficients? If $4.29 is saved for every dollar spent on a social project, is this fact acknowledged, or does the article stop at "significance"? Are you told why the fitted result matters in substance? As Wallis and Roberts said, the economist needs to "appraise its practical importance." Christina Romer does when she writes that "correcting for inventory movements reduces the discrepancy . . . *by approximately half.* This suggests that [inventory] movements are *important*" (*AER,* June 1986, 327; italics supplied). Daniel Hamermesh estimates his crucial parameter K and at the first mention of the results says, "The estimates of K are *quite large,* implying that the firm varies employment only in response to *very large* shocks. . . . Consider what an estimate *this large* means" (*AER,* September 1989, 683; italics supplied). Hamermesh's rhetoric is here close to ideal. It gets to the scientific question of what a magnitude means. On the same page he speaks of magni-

tudes being "fairly large," "very important," "small," and "important" *without merging his standards of large and small with statistical significance*. By contrast, Boissier, Knight, and Sabot write, "In both countries, cognitive achievement bears a *highly significant* relationship to educational level," although they move toward a Yes in the questionnaire by adding, "In Kenya, secondary education raises *H* by 11.75 points, or *by 35 percent of the mean*" (*AER*, December 1985, 1026; italics supplied). Later in the same paragraph, though, "significantly positive" and "almost significantly positive" become their only criteria of importance. The article by Bossier et al. is coded on the question as No. Even Hamermesh, a definite Yes, backslides later in his own article, writing, "The *K*-hat for the aggregated data in Table 2 are insignificant," by which he means "R. A. Fisher 'Insignificant' at 5 percent." But he adds, wisely, "and *very small; and the average values of the p-hat are much higher* than the pooled data" (685; italics supplied).

13. *Does the article discuss the scientific conversation within which a coefficient would be judged "large" or "small"?* Christina Romer, again, remarks that "The existence of the stylized fact [which is to say, the scientific conversation at its present stage] that the economy has stabilized implies a general consensus" (*AER*, June 1986, 322). She then tests it in non-Fisherian ways. Gosset, we've noted, did not put weight on a significance test considered in isolation. Science is social. His practice of repeating experiments, comparative study, and estimation of upper and lower bounds on "real error" contributed, we have noted, to Guinness's—and Gosset's—fortune (Ziliak 2007).

14. *Does the article refrain from choosing variables for inclusion in its equations solely on the basis of statistical "significance"?* Again a No means that the canned program, à la Fisher's *Statistical Methods of Research Workers*, governs the argument. Stepwise regression, as used in urban economics and epidemiology, indulges this fallacious drop-and-add procedure. But there is no scientific reason—unless a reason is provided, and it seldom is—to drop an "insignificant" variable. If the variable is important substantively but is dropped from the regression because it is Fisher-insignificant, the resulting fitted equation will be misspecified, which is to say that the statistical experiment will be incorrectly controlled. Only by the irrelevant standard of producing the most *statistically* significant results will such stepwise regressions make sense. If an equation explaining housing construction drops the price of housing because it was found in an earlier

regression to be Fisher-insignificant, the remaining variables will have to take on the price effect—an effect an economist would believe to be important whether or not it shows with high likelihood through the variability of the sample at hand. She can test her belief in the price effect by looking at the magnitudes, using, for example, the highly advanced technique common in data-heavy articles in physics journals: "interocular trauma." That is, she can look and see if the result hits her between the eyes.

15. *Later, after the crescendo, does the article refrain from using statistical significance as the criterion of scientific importance?* Sometimes the referees will have insisted unthinkingly on a significance test, and the appropriate t's and F's (and in fields other than economics, which does not use them much, R^2's) have therefore been inserted. Even our friend the late Jack Hirshleifer, a distinguished fellow of the American Economic Association, was forced to insert t-statistics into a published article—though Jack himself was thoroughly persuaded by Bayes's rule, loss functions, and economic significance (2004). A Yes here means the author depends on other and more scientifically relevant arguments such as simulations of the magnitudes.

16. *Is statistical significance portrayed as decisive, a conversation stopper, conveying a sense of an ending?* Christine Romer (*AER,* June 1986) and Jeffrey Sachs (*AER,* March 1980) both misuse statistical significance, but in neither article does such a "finding" dominate the empirical work. In Michael Darby's article, by contrast, the misuse is significantly more pervasive (*AER,* June 1984, 311, 315).

17. *Does the article ever use an independent simulation—as against a use of the regression coefficients as inputs into further calculations—to determine whether the coefficients are reasonable?* Blomquist, Berger, and Hoehn (*AER,* March 1988), whom we have criticized, use simulations, but the statistical significance of regression coefficients dictates their inputs. They simulate the ranking of cities by amenities and produce a dollar figure for the differential between the worst and best. Their differential between best and worst, by the way, seems low at $5,146 (96). At that price, one student loan could get you out of downtown Detroit and into sunny Santa Barbara. Perhaps the more important determinants of amenities were mistakenly omitted from the simulation because they had proven "insignificant."

18. *In the concluding sections is statistical significance separated from policy, economic, or scientific significance?* In medicine and epidemiology

and especially psychology the concluding sections are often sizeless summaries of significance tests reported earlier in the article. Significance this, significant that. In economics, too.

19. *Does the article use the word* significant *unambiguously?* Darby (*AER,* June 1984, 310) scores on this, as on many other questions, a No. "First," he writes, "we wish to test whether oil prices, price controls, or both have a significant influence on productivity growth." Like the authors of the articles on "critical thinking" and swineherd "salmonella," he is merging two meanings of *significant,* namely, "a partial account of probability dictated by R. A. Fisher" and "large or small enough in effect to matter for policy or science."

To repeat, we do not pretend that our nineteen-item questionnaire provides a complete description of best-practice empirical economics, only that it gets to the main points of *how much* and *who cares?* We failed, for example, to ask about "confidence intervals" along the lines of a question 20: *Does the article report confidence intervals, using them to interpret economic significance and not merely as a substitute for pointwise statistical significance?* Had we asked this question about confidence intervals of every article in the *AER*—and, again, we wish we had—we predict that in the 1980s the percentage Yes would have been less than 5 percent. As Zellner (1984, 277–80) discovered in his survey of best practice long ago, economists do not think substantively about the meaning of confidence intervals and don't anyway report them. The advantage of confidence intervals is that they draw attention to the magnitudes. And, though we did not measure article by article the incidence of the fallacy of the transposed conditional—an important question 21—any reader who understands the fallacy will believe with the authors that the share of articles indulging it and therefore disseminating the wrong probabilities for their hypotheses is near 100 percent.

6

A Lot Did Go Wrong
in the *American Economic Review*
during the 1980s

*Sophisticated, hurried readers continue to judge works on the so-
phistication of their surfaces. . . . I mean only to utter darkly that
in the present confusion of technical sophistication and significance,
an emperor or two might slip by with no clothes.*

ANNIE DILLARD 1988, 31

Table 6.1 shows the results for the 1980s, arranged from worst result to
best by question. That is,

- Seventy percent of the articles in the *American Economic Review* of the
  1980s made no distinction at all between statistical significance and
  economic or policy significance, what we have called oomph. Look at
  questions 16 and 18.

- At the first use of statistical significance, typically in the "Results" section, 53 percent of the articles considered nothing aside from t- and
  F-statistics (question 7).

- Seventy-two percent did not ask how large is large in the context of
  what other economists or scientists in the field have found (question
  13). They did not ask what standards other scientists have used to determine "importance." Awareness that science is social seemed to
  greatly improve the practice. Of 131 articles that did *not* mention the
  work of other economists as a quantitative context for their own work,
  over three-quarters let statistical significance decide every question of
  substance. Of the 50 articles that *did* mention the work of other economists as a context, only one-fifth let statistical significance decide.

74

- About three-fifths of the articles (the complement of question 19) used the word *significant* equivocally, at one moment meaning "two standard errors removed from the null" and at the next "by the substantive magnitude of the coefficient greatly changing our scientific opinion."

TABLE 6.1. The *American Economic Review* of the 1980s Contained Numerous Errors in the Use of Statistical Significance Measured by Percent Yes

Survey Question	Number of Applicable Articles	Percent Yes
Does the article . . .		
8. Consider the power of the test?	182	4.4
6. Eschew reporting all standard errors, *t*-, *p*-, and *F*-statistics, when such information is irrelevant?	181	8.3
17. Do a simulation to determine whether the coefficients are reasonable?	179	13.2
9. Examine the power function?[a]	12	16.7
13. Discuss the scientific conversation within which a coefficient would be judged large or small?	181	28.0
16. Consider more than statistical significance decisive in an empirical argument?	182	29.7
18. In the conclusions, distinguish between statistical and economic significance?	181	30.1
2. Report descriptive statistics for regression variables?	178	32.4
15. Use other criteria of importance besides statistical significance after the crescendo?	182	40.7
19. Avoid using the word *significance* in ambiguous ways?	180	41.2
5. Carefully interpret the theoretical meaning of the coefficients? For example, does it pay attention to the details of the units of measurement and to the limitations of the data?	181	44.5
11. Eschew "sign econometrics," remarking on the sign but not the size of the coefficient?	181	46.7
7. At its first use, consider statistical significance to be one among other criteria of importance?	182	47.3
3. Report coefficients in elasticities or in some other useful form that addresses the question of "how large is large"?	173	66.5
14. Avoid choosing variables for inclusion solely on the basis of statistical significance?	180	68.1
10. Eschew "asterisk econometrics," the ranking of coefficients according to the absolute value of the test statistic?	182	74.7
12. Discuss the size of the coefficients?	182	80.2
1. Use a small number of observations, such that statistically significant differences are not found merely by choosing a very large sample?	182	85.7
4. Test the null hypotheses that the authors said were the ones of interest?	180	97.3

Source: All full-length articles that use tests of statistical significance published in the *American Economic Review,* January 1980–December 1989, excluding the *Proceedings* (McCloskey and Ziliak 1996, 105).

Note: "Percent Yes" is the total number of Yes responses divided by the relevant number of articles.

[a]Of the articles that mention the power of a test, this is the fraction that examined the power function or otherwise corrected for power.

Multiple-author articles were a little worse than single-author articles, as you can see in table 6.2 . You can understand the modestly significant fact as an outcome of what economists call the "common-pool" or "fisheries" problem. Each author is likely to take a free ride on the efforts of the others and therefore will, on average, devote less care. "My coauthor made me do it" is what she will say. (For example, McCloskey says it of her Fisherian-egregious 1983 article on medieval storage.)

Faculty at "tier 1" schools did a little better than others but not by much (see table 6.3). It is not the case that the scholars at higher-reputation schools avoid misusing statistical significance—contrary to some irritated dismissals of our point by friends at tier 1 schools. "Those peasants at tier 4 do it. Not we aristocrats." On the contrary, the practice is scandalously bad up and down the academic hierarchy, hardly justifying the invidious terminology of tier 1 in the first place. And tier 1 economists did *worse* than others in practices we do not report explicitly here, namely, using entire universes as "samples" and treating a sample of convenience as a random and representative sample.

Some of the worst examples of erroneous null-hypothesis significance testing occur in finance and macroeconomics. For example, one article offers "an alternative test of the CAPM and report[s] . . . test results that are free from the ambiguity imbedded in the past tests" (*AER*, January 1980, 660). The authors "test" five hypotheses: the intercept differs from zero, the slope differs from zero, the adjusted coefficient of determination is 1.0, the trend of the intercept is zero, and the trend of the adjusted coefficient of determination is zero (664–65). Nowhere is the size of the esti-

TABLE 6.2. Multiple Authors Appear to Have Coordination Problems, Making the Abuses Worse Measured by Percent Yes

Survey Question	Multiple-Author Articles	Single-Author Articles
Does the article . . .		
7. At its first use, consider statistical significance to be one among other criteria of importance?	42.2	53.4
10. Eschew "asterisk econometrics," the ranking of coefficients according to the absolute value of the test statistic?	68.8	79.2
12. Discuss the size of the coefficients?	76.7	84.1
1. Use a small number of observations such that statistically significant differences are not found merely by choosing a very large sample?	77.8	84.8

Note: "Percent Yes" is the total number of Yes responses divided by the relevant number of articles.

mated coefficients discussed. The results are *ranked* according to the number of times the *t*-statistic is greater than 2.0 (667). There is no mention of the only relevant scientific test of a stock market investment model, namely, its ability to make money. The only Yes the article earned was one for specifying the null according to what the theory implied.

An article on social welfare in Boston is a regular traffic jam of significance.

> The statistically significant [read (1) sampling theory "significant" at 5 percent] inequality aversion is in addition to any unequal distribution of inputs resulting from different social welfare weights for different neighborhoods. The KP results allowing for unequal concern yield an estimate of q of -3.4. This estimate is significantly [read (2) some numbers are smaller than others] less than zero, indicating aggregate outcome is not maximized. At the same time, however, there is also significant [read (3) a moral or scientific or policy matter] concern about productivity, as the inequality parameter is significantly [read (4) a joint observation about morality and numbers] greater than the extreme of concern solely with equity. (*AER*, March 1987, 46).

TABLE 6.3. Authors at Tier 1 Departments Do Better than Others in Many Categories . . . but Even Their Practice Is Nothing to Celebrate Measured by Percent Yes

Survey Question	Tier 1 Departments	Other Departments
Does the article		
1. Use a small number of observations, such that statistically significant differences are not found merely by choosing a very large sample?	91.3	83.9
12. Discuss the size of the coefficients?	87.0	78.9
10. Eschew "asterisk econometrics," the ranking of coefficients according to the absolute value of the test statistic?	84.8	71.4
7. At its first use, consider statistical significance to be one among other criteria of importance?	65.5	41.2
5. Carefully interpret the theoretical meaning of the coefficients? For example, does it pay attention to the details of the units of measurement and to the limitations of the data?	60.0	37.5
19. Avoid using the word *significance* in ambiguous ways?	52.4	37.5
18. In the conclusions, distinguish between statistical and economic significance?	50.0	23.1

Note: When McCloskey and Ziliak 1996 went to print the National Research Council had assigned the following schools to tier 1: Chicago, Harvard, MIT, Princeton, Stanford, and Yale.

"Percent Yes" is the total number of Yes responses divided by the relevant number of articles.

There is no mention of the loss function of social policy misapplied. Only—ominously—"KP" results.[1]

Of course some of the economists in the 1980s got it approximately right—though not one of the 182 articles, we repeat, refrained *entirely* from misusing statistical significance. We've mentioned Christine Romer and Daniel Hamermesh as exemplary scholars on such matters. Thus, too, Kim B. Clark: "While the union coefficient in the sales specification is twice the size of its standard error, it is *substantively small;* moreover, with over 4,600 observations, the *power of the evidence* that the effect is different from zero is not overwhelming" (*AER,* December 1984, 912; italics supplied). This is about as sophisticated as it got in the 1980s.

Most of the economic scientists in the decade did not come close to grasping that Fisher-significance is a minor matter of fit, not a major matter of oomph. Of the 70 percent of the articles that flatly mistook statistical significance for economic significance, about 70 percent again failed to report even the magnitudes of influence between the economic variables they investigated. In other words, during the 1980s about one-half of the empirical articles ($\approx 0.7 \times 0.7$) published in the *American Economics Review* did not establish their claims of economic significance.

7

Is Economic Practice Improving?

Because the great controversies of the past often reach into modern science, many current arguments cannot be fully understood unless one understands their history.

ERNST MAYR 1982, 1

Our 1996 article reporting these findings, we repeat, had almost no impact. Horowitz and a few others have told us that they disagree. They see progress and give us credit for causing some of it. That would be nice. But lack of impact is typical in other fields subjected to such criticism, for example, medicine and biology. In economics after 1996 our impression is that the phrase "economically significant" started to turn up a little bit more, and we flatter ourselves that a new if vague awareness that *something* is fishy in the standard error is responsible for this.

Yet in the several score of seminars we have given together and individually on the subject since 1996 we have been told repeatedly, "After the 1980s, the decade you examined in your splendid 1996 article, best practice improved. Things are getting better, partly because of your wonderfully persuasive work. Look, for example, at this *new* Fisherian test on cointegration I have devised."

We are very willing to believe that since the 1980s our colleagues have stopped making an elementary error and especially that we have changed their minds. But being readers of typical economics articles since that first decade of personal computers we seriously doubted that fishing for significance had much abated. Like our critics, we are empirical scientists. And so in a second article, published in 2004, we reapplied the nineteen-item questionnaire employed in our 1996 article to all the full-length empirical articles of the *next* decade of the *AER,* the 1990s. (By the way, we made a terribly embarrassing error in the published article, omitting fully 50 [!] of the 187 articles for the 1990s. We are grateful to Kevin Hoover and Mark Siegler for pointing this out [Hoover and Siegler 2008]. When we included the missing articles, however, essentially nothing changed—as, indeed, one would expect if sampling theory is correct unless our initial erroneous selection of articles was biased. Apparently it was not.)

We found that the standard error is *not* getting smaller. By the standard of citizen-in-the-street common sense and of advanced decision theory it is getting worse. Of the 187 relevant articles published in the 1990s, 79 percent mistook statistically significant coefficients for economically significant coefficients (as against 70 percent in the earlier decade). In the 1980s a disturbing 53 percent had relied exclusively on statistical significance as a criterion of importance at its first use; in the 1990s an even more disturbing 60 percent did.

Table 7.1 reports the results, distinguished by decades, for all of the 369 full-length articles that test for significance and published in the *AER* from January 1980 to December 1999 (cf. Ziliak and McCloskey 2004a, 2008). We have at hand, to repeat, the whole population, not a sample. The urn of nature is poured out before us. Unlike most of our colleagues in economics, therefore, we will refrain from calculating statistics relevant only to inference from *samples* to a population, such as an imaginary "statistical significance" of the differences between the two decades.

Like table 6.1, we here rank in ascending order each item of the questionnaire according to "Percent Yes" (in the 1990s). A Yes means, remember, that the article took what most major statistical theorists since Edgeworth have regarded as the correct action on the matter. For example, in the 1980s, 4.4 percent of the articles considered the power of the tests. (We do not believe it is accidental that every article that considered power also considered "a quantitative criterion of what is to be considered a large departure.") That is, 4.4 percent did the correct thing by considering also the probability of a Type II error. In the 1990s, 8 percent did. An encouraging trend. But still only about 8 percent of the articles.

The change in practice is isolated in tables 7.2 and 7.3. In the 1980s only 44.5 percent of the articles paid careful attention to the theoretical and accounting meaning of the regression coefficients (question 5). That is, in the 1980s the reader of an empirical article in the *AER* was nearly six times out of ten left wondering how to interpret the *economic* meaning of the

TABLE 7.1. The *American Economic Review* Had Numerous Errors in the Use of Statistical Significance, 1980–99

Survey Question	Percent Yes in 1990s	Percent Yes in 1980s
Does the article . . .		
8. Consider the power of the test?	8.0	4.4
6. Eschew reporting all standard errors, *t*-, *p*-, and *F*-statistics, when such information is irrelevant?	9.6	8.3
16. Consider more than statistical significance decisive in an empirical argument?	20.9	29.7
11. Eschew "sign econometrics," remarking on the sign but not the size of the coefficient?	21.9	46.7
14. Avoid choosing variables for inclusion solely on the basis of statistical significance?	27.3	68.1
15. Use other criteria of importance besides statistical significance after the crescendo?	27.8	40.7
10. Eschew "asterisk econometrics," the ranking of coefficients according to the absolute value of the test statistic?	31.0	74.7
17. Do a simulation to determine whether the coefficients are reasonable?	32.6	13.2
19. Avoid using the word *significance* in ambiguous ways?	37.4	41.2
7. At its first use, consider statistical significance to be one among other criteria of importance?	39.6	47.3
9. Examine the power function?[a]	44.0	16.7
13. Discuss the scientific conversation within which a coefficient would be judged large or small?	53.5	28.0
18. In the conclusions, distinguish between statistical and economic significance?	56.7	30.1
2. Report descriptive statistics for regression variables?	66.3	32.4
1. Use a small number of observations, such that statistically significant differences are not found merely by choosing a very large sample?	71.1	85.7
12. Discuss the size of the coefficients?	78.1	80.2
5. Carefully interpret the theoretical meaning of the coefficients? For example, does it pay attention to the details of the units of measurement and to the limitations of the data?	81.0	44.5
4. Test the null hypotheses that the authors said were the ones of interest?	83.9	97.3
3. Report coefficients in elasticities, or in some other useful form that addresses the question of "how large is large"?	86.9	66.5

Source: All full-length articles that use tests of statistical significance published in the *American Economic Review* in the 1980s ($N = 182$) and 1990s ($N = 187$; Ziliak and McCloskey 2004a). Table 1 in McCloskey and Ziliak 1996 reports a small number of articles for which some questions in the survey do not apply.

Note: "Percent Yes" is the total number of Yes responses divided by the relevant number of articles.

[a]Of the articles that mention the power of a test, this is the fraction that examined the power function or otherwise corrected for power.

coefficients. In the 1990s the share taking the correct action rose to 81 percent, a net improvement of about 36 percentage points. "Economic significance" appeared increasingly alongside "statistical significance." This is one meaning of oomph: a big change, important for the science.

Similarly, the percentage of articles reporting units and descriptive statistics for regression variables of interest rose by 34 percentage points from 32.4 percent to 66.3 percent (question 2). Excellent. Perhaps after the initial dazzlement with what could be done by way of regression analysis on one's personal computer the economists paused to think about *what* they were doing. We don't claim to know for certain. Some other margins improved. Gains of more than 20 percentage points were made in, for example, the share of articles discussing the scientific conversation in which a coefficient would be judged large or small, the share of articles keeping statistical and economic significance distinct in the "conclusions" section, and the share of articles doing a simulation to determine whether the estimated coefficients are reasonable. Our definition of *simulation* is broad. It includes articles that check the plausibility of the regression results by making, for example, Harberger-Triangle-type calculations on the

TABLE 7.2. The Economic Significance of the *American Economic Review* Has in Some Regards Improved (measured by net percentage point difference, 1980–99)

Survey Question	Percent Yes in 1990s	Net Improvement since 1980s
Does the article . . .		
5. Carefully interpret the theoretical meaning of the coefficients? For example, does it pay attention to the details of the units of measurement, and to the limitations of the data?	81.0	+36.5
2. Report descriptive statistics for regression variables?	66.3	+34.0
9. Examine the power function?[a]	44.0	+27.3
18. In the conclusions, distinguish between statistical and economic significance?	56.7	+26.6
13. Discuss the scientific conversation within which a coefficient would be judged large or small?	53.5	+25.5
17. Do a simulation to determine whether the coefficients are reasonable?	32.6	+19.4

Source: All full-length articles that use tests of statistical significance published in the *American Economic Review* in the 1980s (N = 182) and 1990s (N = 187; Ziliak and McCloskey 2004a). Table 1 in McCloskey and Ziliak 1996 reports a small number of articles for which some questions in the survey do not apply.

Note: "Percent Yes" is the total number of Yes responses divided by the relevant number of articles.

[a]Of the articles that mention the power of a test, this is the fraction that examined the power function or otherwise corrected for power.

basis of descriptive data. But the article that uses statistical significance as the sole criterion for including a coefficient in a later simulation is coded No, which is to say that it does not do a simulation to determine whether the coefficient itself is reasonable.

Such gains are commendable. But the room for additional gain is immense. For example, as we said, that 8 rather than 4 percent consider power is nice, but this still leaves 92 percent of the articles risking high levels of a Type II error. For most questions the improved levels of performance are less than impressive. Only perhaps in the matter of question 5—concerning the interpretation of theoretical coefficients in the Goldberger way—does the improvement approach levels that most statisticians would agree are good practice. Thus the sorites.

For example, in the 1990s *two-thirds* of the articles did not make calculations to determine whether the estimated magnitude of the coefficients made sense (question 17)—only a third, we found, had simulated the

TABLE 7.3. But the Essential Confusion of Statistical and Economic Significance is Getting Worse (measured by net percentage difference, 1980–99)

Survey Question	Percent Yes in 1990s	New Percentage Point Change since 1980s
Does the article . . .		
10. Eschew "asterisk econometrics," the ranking of co-efficients according to the absolute value of the test statistic?	31.0	−43.7
14. Avoid choosing variables for inclusion solely on the basis of statistical significance?	27.3	−40.8
11. Eschew "sign econometrics," remarking on the sign but not the size of the coefficient?	21.9	−24.8
1. Use a small number of observations, such that statistically significant differences are not found merely by choosing a very large sample?	71.1	−14.6
4. Test the null hypotheses that the authors said were the ones of interest?	83.9	−13.4
15. Use other criteria of importance besides statistical significance after the crescendo?	27.8	−12.9
16. Consider more than statistical significance decisive in an empirical argument?	20.9	−8.8
7. At its first use, consider statistical significance to be one among other criteria of importance?	39.6	−7.7

Source: All full-length articles that use tests of statistical significance published in the *American Economic Review* in the 1980s (N = 182) and 1990s (N = 187; Ziliak and McCloskey 2004a). Table 1 in McCloskey and Ziliak 1996 reports a small number of articles for which some questions in the survey do not apply.

Note: "Percent Yes" is the total number of Yes responses divided by the relevant number of articles.

effect of their coefficients with the analytical force at least of an introductory economics text. Skepticism about alleged effects, tested by simulation, is, by contrast, normal practice in cell biology and most branches of physics.

An examination of many years of the *Physical Review* revealed no tests of statistical significance. Well, almost none by the prolix standard of economics (McCloskey and Ziliak 2008). So, too, in *Cell* and also in *Science* if one considers only the biologists and chemists and geologists publishing there. Zellner notes that Harold Jeffreys was a famous physicist, geologist, mathematician, probability theorist, philosopher, and applied scientist— but he did not and would not use the title, "statistician." His titles at Cambridge University were "Reader in Geophysics" and "Plumian Professor of Astronomy and Experimental Philosophy." As Zellner sees it, Jeffreys "wrote his books *Scientific Inference* and *Probability Theory* [Theory of Probability (1939)] to instruct his fellow scientists with respect to the philosophy of science and how to analyze their data appropriately." "Note," Zellner continues, "that he used the title *Probability Theory, not* Statistics, since for many physicists Statistics is a bad word. I think that it was Lord Rutherford who said that if you need statistics to analyze your experimental data, you better redesign your experiment. . . . While the methods are 'statistical,' most physicists would not like to be called 'statisticians'."[1]

This is not to say, we repeat, that physicists and biologists and chemists and geologists never, ever use the tests. Indeed, long before Jeffreys, astrophysicists, geophysicists, and research chemists had made seminal contributions to the theory of statistics and certainly helped to spread it. In 1898 Sydney Lupton published *Notes on Observations, Being an Outline of the Methods Used for Determining the Meaning and Value of Quantitative Observations and Experiments in Physics and Chemistry, and for Reducing the Results Obtained* (Lupton 1898). Lupton endeavored primarily to introduce chemists and physicists to the "method of least squares" (i). Gosset himself, who was, as we have said, trained at Oxford in chemistry, received some early instruction in "the theory of errors" from Lupton. He was particularly helped by a 120-page book by G. B. Airy (1801–92), the Astronomer Royal and inventor of, among other things, the instrument used to establish the prime meridian at Greenwich.[2] But the test of significance, then as now, was relegated to the corners of the chemical and physical sciences.

A good thing, too. In 1946 an entomologist, the young E. O. Wilson, published a hostile review of Haavelmo's 1944 "The Probability Approach in Econometrics." Wilson's message was: fields that don't use statistical significance nonetheless advance.[3] Faster. Many biologists reporting their results in *Science* nowadays are less clear minded than the physicists are on the matter. And in their own journals the medical scientists, like the social scientists and some biologists, are very confused indeed.

One additional problem, which is often somewhat carelessly taken to be our main objection—it is not, though it is worrisome enough on its own—is that statistical *insignificance* is nonpublic, decided behind the curtain where the emperor sits. In the 1990s three fourths of the articles in the *American Economic Review* chose variables *for inclusion* (i.e., pretests) solely on the basis of statistical significance, a net decline in best practice of fully 43 percentage points (question 14). As Kruskal put it in his 1968 article:

> Negative results are not so likely to reach publication as are positive ones. In most significance-testing situations a negative result is a result that is not statistically significant, and hence one sees in published articles and books many more statistically significant results than might be expected. . . . The effect of this is to change the interpretation of published significance tests in a way that is hard to analyze quantitatively. (1968a, 245)

The response to question 14 shows that economists made it hard in the 1990s to analyze quantitatively, in Kruskal's sense, the real world relevance of their "significant" results. It's the problem of insisting on significance, encouraged by the incentives to publish—as numerous economists have noted in cynical amusement or despairing indignation.

The core confusion over the meaning of significance testing, alas, has gotten worse, as shown in table 7.3. For example, fully 78 percent of the articles in the 1990s engaged in what we call "sign econometrics" (question 11). In the 1980s, 53 percent did, troubling enough. Sign econometrics is worse for scientific thinking when the economist does not report confidence intervals. Perhaps because they are not often trained in the error-regarding traditions of engineering or chemistry, economists seldom report confidence intervals. Thus Hendricks and Porter, in "The Timing and Incidence of Exploratory Drilling on Offshore Wildcat Tracts" (1996, 404), write, "In the first year of the lease term, the coefficient of HERF is positive, but not significant. This is consistent with asymmetries of lease holdings mitigating any information externalities and enhancing coordination, and therefore reducing any incentive to delay." "HERF" is Hendricks and Porter's Herfindahl index of the dispersion of leaseholdings among bidders at auction. Yet the reader is nowhere told how much HERF changed the probability the winners would engage in exploratory oil drilling.

Over two-thirds of the articles in the 1990s *ranked* the importance of their estimates according to the absolute values of the test statistics, ignoring the estimated size of the economic impact (question 10). In other words, asterisk econometrics became in the 1990s a good deal more popular in economics (it has long been popular in psychology and sociology), increasing over the previous decade by 44 percentage points. Bernanke and Blinder (1992, 905, 909), Bernheim and Wantz (1995, 547), and Kachelmeier and Shehata (1992, 1130), for example, published tables featuring a hierarchy of p-, F-, and t-statistics, the totems of asterisk econometrics. Two decades ago Zellner pointed out that a sample of eighteen articles in 1978 never had "a discussion of the relation between choice of significance levels and sample size" (one version of the problem we emphasize here) and usually did not discuss *how far* from 5 percent the test statistic was. Zellner remarked coolly that "there is room for improvement in analysis of hypotheses in economics and econometrics" (1984, 277–80). It would seem so.

The econometrician Jeffrey Wooldridge (2004) agrees strongly with our claim that statistical significance is not a necessary or sufficient condition for economic significance, and he makes the point in his fine introductory textbook on econometrics (2000). Wooldridge wrote in 2004, "I attend too many empirical workshops where the sizes of the coefficients are not discussed"—which is simply more evidence, he believes, "that econometric practice may indeed be in trouble" (577). He is certainly correct to caution, however, that some researchers, as he puts it, "push" magnitudes of *economic* significance in ways they should not. His is an assertion about the ethics of communication in science and public affairs that we find both poignant and understudied.

But even Wooldridge balks. "Pushing" an economically large *though noisily estimated* effect is not a misuse—or a "stretch," as he says—of professional ethics. It is precisely the ethical thing to do. To argue otherwise is to fall into the mistaken belief that statistical significance *can* provide a screen through which the results can be put, to be examined then for *substantive* significance if they make it through the significance screen. This is a superstition, though widely believed. Statistical significance is *not* necessary for a coefficient to have substantive significance and therefore *cannot* be a suitable prescreen. As we showed in our 1996 article, the noisily estimated benefit-cost ratio of about 4:1 in the state of Illinois unemployment insurance program is one such instance. The loss of jobs and wages attendant to the action *no change in employment policy*—which is what the mechanical rule of statistical significance suggested—we find disturbing. It would be no "stretch" at all to bring it to the attention of economists. A similar point could be made about the failure we have mentioned of the U.S. Department of Labor to release certain black urban unemployment rates.

And we do not agree with Wooldridge's claim that we "oversell" the extent and error of sign and asterisk econometrics. He cites Bernheim and Wantz (1995)—an article that we would say exemplifies the very problems both Wooldridge and we are complaining about—and defends them on the grounds that "while the coefficients have the signs they expect from theory, they are not willing to claim additional support for their theory because the effects are statistically insignificant" (2004, 579). The problem we see—and we believe Wooldridge will on reflection agree with us—is that claiming additional support for the negative signs because the effects are statistically significant is invalid. A tightly fit and negatively signed coefficient on bond yield may be in effect, for economic purposes, zero, that is, substantively insignificant. As we've argued at length, and as Wooldridge (2000) suggests in his textbook, sign without size, and sign without size without confidence intervals, and sign without size without confidence intervals without loss functions, is simply beside the point.

One more asterisk
To rest like eyes of dead fish—
Rigor mortis stars.

Table 7.4 shows what an article looks like in other respects if statistical significance is the only criterion of importance in the article at first use. Of 137 full-length articles published in the 1990s (the inadvertent selection

TABLE 7.4. If Only Statistical Significance Is Said to Be of Importance at Its First Use (Question 7), Then Statistical Significance Tends to Decide the Entire Argument

		Does Not Consider the Test Decisive (Question 16)		
		No	Yes	Total
Considers more than the test at the first use (Question 7)	No	80	7	87
	Yes	32	18	50
	Total	112	25	137

Source: 137 of 187 full-length articles published in the *American Economic Review* in the 1990s and analyzed in Ziliak and McCloskey 2004a.

Note: No means "did the wrong thing." Yes means "did the right thing." In the 1980s data the first row was 86–10–96.

studied in our 2004 article), 80 made both mistakes (question 7 = No and question 16 = No). Looking at the data in our 2004 articles, of the 87 articles using only statistical significance as a criterion of importance at first use, fully 80 of the 87 considered statistical significance the last word. Cross tabulations on the 1980s data reveal a similar though slightly better record.

What is most distressing, however, is the increasing conflation of statistical and economic significance, indicated by the responses to questions 16 and 7. The main findings are:

- Seventy-nine percent of the 187 empirical articles published in the 1990s in the *American Economic Review* did not distinguish statistical significance from economic significance (question 16). In the 1980s, 70 percent did not—scandalous enough.

- At the first use of statistical significance, typically in the "Estimation" or "Results" section of an article, 60 percent in the 1990s did not consider *anything but* the size of the test statistics as a criterion for the inclusion of variables in future work. In the 1980s, 53 percent (question 7)—seven percentage points fewer—did so.

8

How Big Is Big in Economics?

The art of designing all experiments lies even more in arranging matters so that ρ [the correlation coefficient] is as large as possible than in reducing σ_x^2 and σ_y^2 [the variance].

STUDENT 1923, 271

Of course, not everyone gets everything wrong. The *American Economic Review* over the past two decades has been filled with superb economic science. In our opinion a good fraction of all the articles we read can be described this way—even though a supermajority make significant mistakes in the use of statistical significance. In other words, we do *not* accept the opinion of one eminent econometrician we consulted, who dismissed our case by remarking cynically that after all such idiocy as we had found was to be regarded as par for the course in such a contemptible publication as the *American Economic Review*. Nor do we agree with a former editor of the *Review*, who said that he "basically agreed" with our criticism of statistical significance but then added that "Young people have to have careers" and so the abuse should continue.

In the survey of the articles published in the 1990s we decided that we should show the names of the economists with their scores. It seemed to be the only way to prevent the reply, at any rate from the very authors surveyed, that "only idiots do that." These are not idiots, even those, such as Chairman Bernanke, who seem not to grasp that Fisherian significance is not the same as oomph. They are a selection of the best economic scientists of the 1990s, and no one who reads their work with understanding could seriously claim otherwise. Yet we found that even Nobel Prize winners continue to commit what most careful students of

89

the matter would consider grammar school errors in the use of statistical significance.

Table 8.1 reports the author rankings by economic significance in five brackets. If an article chose between fifteen and nineteen actions correctly, as Gary Solon's article did (1992), then it is in the top bracket, the best if not perfect practice. We note with pleasure, by the way, that economists (Goldin, Allen) from our own field of economic history wrote two of the eight articles in the top bracket—and the Zimmerman, Solon, and Craig and Pencavel articles can be easily described as studies in labor history and the history of ideas, meaning that five of the eight articles in the top category are historical. No surprise, we say. If, on the other hand, an article chose only between six and eight actions correctly, as Gary Becker, Michael Grossman, and Kevin Murphy did (1994), then it is in the fourth or "poor" bracket, second to last.

There's faint evidence in these data of what psychologists call the "reliability" of the survey, namely, that the same author making an appearance in different articles tends to score roughly the same on both occasions (or in one case—Angrist's 3 scores of Good—on all occasions). "Faint," we said. Three Exemplary authors, for example, reappeared in a lower category—Good, Poor, or Very Poor—though, unhappily for the power of such a test, no others from the Exemplary category reappeared at all. In 20 of 39 cases of double appearances in the 1990s ($N = 187$) an author spanned more than one rank; in other words, about half the time. We did not consider articles by authors making 3 or more appearances. Nor did we make any adjustment for double appearances that involve coauthors, or a different set of coauthors, in relation to single-author articles. We suspect that if one did make these and other kinds of adjustments—such as "author maturity," however defined—the frequency of rank change would diminish, pushing the distribution closer toward the perfect diagonal.

The null hypothesis of zero reliability would imply a flat distribution across successive ranks. Table 8.2 shows the actual distribution. Consistent authors would produce only diagonal entries: Exemplary-Exemplary, Good-Good, Fair-Fair, and so forth. In reality a little less than a third were consistent, that is, "on the diagonal." The either/or rhetoric of significance, coupled with the common pool problems associated with multiple authorship, enables such inconsistency. That is one explanation for the scattered ranks. But, to make again the central point of our book, *how far is far* from flat or diagonal?

TABLE 8.1. Author Rankings of the 1990s by Economic Significance Measured by Percent Yes, That Is, Good, in the Nineteen-Question Survey (year and month of publication in brackets)

15–19 Yes Answers: Exemplary
Solon [6/92]
Zimmerman [6/92]
Goldin [9/91]
Craig and Pencavel [12/92]
Anderson and Holt [12/97]
Gali [12/91]
Hercowitz and Sampson [12/91]
Ransom [3/93]
Ciccone and Hall [1/96]
Allen [3/92]
Davis and Haltiwanger [12/99]
Ausubel [12/90]

12–14: Good
Simon [12/98]
Angrist and Evans [6/98]
Berk, Hughson, and Vandezande [9/96]
Myagkov and Plott [12/97]
Gordon and Bovenberg [12/96]
Angrist [12/95]
Gilligan [12/92]
Hoover and Sheffrin [3/92]
Benhabib and Jovanovic [3/91]
Angrist [6/90]
Cecchetti, Lam, and Mark [6/90]
Baker and Benjamin [9/97]
Paxson [3/92]
Blank [12/91]
Froot and Obstfeld [12/91]
Wilcox [9/92]
Cogley and Nason [6/95]
Pesando [12/93]
Watson [3/94]
Del Boca and Flynn [12/95]
Wolfram [9/99]
Engel and Hamilton [9/90]
Ghosh and Masson [6/91]
Cooper and Haltiwanger [6/93]
Kashyap and Wilcox [6/93]
Banks, Blundell, and Tanner [9/98]

9–11: Fair
Brainerd [12/98]
Calomiris and Mason [12/97]
Morrison and Schwartz [12/96]
Landers, Rebitzer, and Taylor [6/96]
Guiso, Jappell, and Terlizzese [3/96]
Borjas [6/95]
Kaminsky [6/93]
Calvo and Leiderman [3/92]

Fair and Shiller [6/90]
Sauer and Leffler [3/90]
Schachar and Nalebuff [6/99]
Craft [12/98]
Dyck [9/97]
Genesove and Meyer [6/97]
Pontiff [3/97]
Rosenszweig and Wolpin [12/94]
Currie and McConnell [9/91]
Hendry and Ericsson [3/91]
Pitt, Rosenzwieg, and Hassan [12/90]
Berry, Levinsohn, and Pakes [6/99]
Yano and Nugent [6/99]
Ham, Sveinar, and Terrell [12/98]
Hallock [9/98]
Rajan and Zingales [6/98]
Ichnowski, Shaw, and Prennushi [6/97]
Nalbantian and Schotter [6/97]
Wilhelm [9/96]
Fuchs [3/96]
Rotemberg and Woodford [3/96]
Griliches and Cockburn [12/94]
James [9/93]
Forsythe, Nelson, Neumann, and Wright [12/92]
Stratman [12/92]
Lin [3/92]
Viscusi and Evans [6/90]
King, Plosser, Stock, and Watson [6/91]
Hamilton [3/92]
Coleman [3/96]
Christiano and Eichenbaum [6/92]
Metrick [3/95]
Bakshi and Chen [3/96]
Evans, Farrelly, and Montgomery [9/99]

6–8: Poor
Mendelson, Nordhaus, Shaw [9/94]
Fernald [6/99]
Gali [3/99]
Murray, Evans, and Schwab [9/98]
Alesina and Peroti [12/97]
Harrigan [9/97]
Dorwick and Quiggin [3/97]
Chevalier and Scharfstein [9/96]
Levin, Kagel, and Richard [6/96]
Trefler [12/95]
Felstein [6/95]
Mark [3/95]
Ashenfelter and Krueger [12/94]
Gale and Scholz [12/94]

(continues)

TABLE 8.1—Continued

Cohen [6/93]
Altonji, Hayashi, and Kotlikoff [12/92]
Bernanke and Blinder [9/92]
Card [9/90]
Aitken and Harrison [6/99]
Levine and Zervos [6/98]
Blonigen [6/97]
Hines [12/96]
Henderson [9/96]
Laitner and Juster [9/96]
Grinblatt, Titman, and Wermers [12/95]
Lemieux, Fortin, and Frechette [3/94]
Hanes [9/93]
Blundell, Pashardes, and Weber [6/93]
Kachelmeier and Shehata [12/92]
Wolff [6/91]
Hardouvelis [9/90]
Wright [9/90]
Card and Krueger [9/94]
Burman and Randolph [9/94]
Palfrey and Prisbrey [12/97]
Peek and Rosengren [9/97]
Levitt [6/97]
Cardia [3/97]
Hamilton [3/97]
Foster and Rosenzweig [9/96]
Hendricks and Porter [6/96]
Ayers and Siegelman [6/95]
Jones and Kato [6/95]
Meyer, Viscusi, and Durbin [6/95]
Fuhrer and Moore [3/95]
Shea [3/95]
Becker, Grossman, and Murphy [6/94]
Persson and Tabellini [6/94]
Alogoskoufis and Smith [12/91]
Fair and Dominguez [12/91]
Bohn [12/90]
Cecchetti [3/92]
Feenstra [3/94]
Nousair, Plott, Riezman [6/95]
Lewis [9/95]
Burnside and Eichenbaum [12/96]
Milyo and Waldfogel [12/99]
Burda and Gerlach [12/92]
Knetter [6/93]
Gruber [6/94]
Bizjak and Coles [6/95]
Cason [9/95]

Cawley and Philipson [9/99]
Leitch and Tanner [6/91]
Friedman and Kuttner [6/92]
Lewbel [9/94]
Gu and Kuhn [6/98]
Cooper, Kagel, Lo and Gu [9/99]
Goldberg and Maggi [12/99]

6: Very Poor
Frankel and Romer [6/99]
Krozner and Stratman [12/98]
Bernard and Jones [12/96]
Munnell, Tootell, Browne, and
    McEneany [3/96]
Attanasio and Browning [12/95]
Marin and Schnitzer [12/95]
Chevalier [6/95]
Currie and Thomas [6/95]
Bronars and Grogger [12/94]
Krozner and Rajan [9/94]
Kim and Singal [6/93]
Bronars and Deere [3/93]
Kashyap, Stein, and Wilcox [3/93]
Falvey and Gemmell [12/91]
Keeley [12/90]
Ramey and Ramey [12/95]
Hamermesh and Biddle [12/94]
Keane [9/93]
Grossman [9/92]
Cukierman, Edwards, and Tabellini
    [6/92]
Wolak and Kolstad [6/91]
Keane and Runkle [9/90]
Roberts and Tybout [9/97]
Engel and Rogers [12/96]
Besley and Case [3/95]
Levine and Renelt [9/92]
Trejo [9/91]
Brainard [9/97]
Bernheim and Wantz [6/95]
Ito [6/90]
Stavins and Jaffe [6/90]
Hallman, Porter, and Small [6/91]
Cecchetti, Kayshap, and Mark [12/97]
Levin and Stephan [3/91]
Lewbel [6/96]
Duffy and Ochs [9/99]
Dahl and Ransom [9/99]

Source: All full length articles that use tests of statistical significance published in the *American Economic Review,* January 1980–December 1989, excluding the *Proceedings* (McCloskey and Ziliak 1996, 105).

Note: "Percent Yes" is the total number of Yes responses divided by the relevant number of articles.

THE BIG SO WHAT? F1 AND F2

But we should ask of our own findings what we ask of our colleagues' findings, that most terrifying of all seminar questions: so what? What does it *matter* that first-rate economists are confusing statistical with substantive significance?

The economist and statistician Milton Friedman (1912–2006) was from 1943 to 1945 associate director of the Statistical Research Group of the Division of War Research at Columbia University. The director of the Research Group, Milton's boss, was Harold Hotelling, by then a famous and influential mathematical statistician and Fisher protege. Hotelling and Friedman's group included other statistical giants such as Kruskal, Wald, Wallis, and Jacob Wolfowitz. They were in close contact with Samuel Wilks, too, a one-time Hotelling student from Columbia who directed a parallel Statistical Research Group at Princeton. Friedman was, statistically speaking, among the giants. He was named a fellow of the American Statistical Association, the Econometric Society, and the Institute of Mathematical Statistics—a triple crown. There is still a nonparametric test named after him, a test he devised in 1937, an important turning point in distribution-free analysis. Independent of Neyman, Friedman pioneered techniques of sequential sampling. And the great statistician and probabilist, Jimmie Savage, himself considered Friedman a statistical "mentor" (Savage 1971a, 441–42).

Listen to Friedman tell of his experience during the war with statistical versus substantive significance.

> One project for which we provided statistical assistance was the development of high-temperature alloys for use as the lining of jet engines and as blades of turbo superchargers—alloys mostly made of chrome, nickel, and

TABLE 8.2. How Consistent Are the Authors of the *American Economic Review?*

If a Repeating Author Scored on One Occasion as Well As	What Was the Level of His or Her *Lower or Equal* Score on the Other Occasion?				
	Exemplary	Good	Fair	Poor	Very Poor
Exemplary	1	1		1	
Good		2	2	4	1
Fair		1	6	4	
Poor			4	7	
Very Poor				5	

Source: Table 8.1.

other metals. . . . Raising the temperature a bit increases substantially the efficiency of the turbine, turbo supercharger, or jet engine. . . . I computed a multiple regression from a substantial body of data relating the strength of an alloy at various temperatures to its composition. My hope was that I could use the equations that I fitted to the data to determine the composition that would give the best result. On paper, my results were splendid. The equations fitted very well [fitted *statistically*, with high R^2] and they suggested that a hitherto untried alloy would be far stronger than any existing alloy. . . . The best of the alloys at that time were breaking at about ten or twenty hours; my equations predicted that the new alloys would last some two hundred hours. Really astounding results! . . . So I phoned the metallurgist we were working with at MIT and asked him to cook up a couple of alloys according to my specifications and test them. I had enough confidence in my equations to call them F1 and F2 but not enough to tell the metallurgist what breaking time the equations predicted. That caution proved wise, because the first one of those alloys broke in about two hours and the second one in about three. (1985, 48–49)

Friedman learned in the 1940s that statistical significance is not the same as substantive, metallurgical significance. You can't derive metallurgical facts and propositions from the facts and propositions of statistical significance only. Speaking in terms of the ancient argument from *sorites* Friedman and his team knew a *fast* from *slow* "breaking time." And if F1 and F2 had not been tested by the metallurgists independent of the regression equation—in economics one might think of policy simulations, calibrated independently, of any tests of significance—and if the *statistical* significance of their qualities had been relied on in the making of airplane wings, thousands of airmen would have died. *That* is why our point matters. That's the answer to So What?

Here's another example. The notion that a nation's prices, let us say American prices, are determined by the exchange rate and prices in the world is called "purchasing power parity." The usual test of whether purchasing power parity is true will regress prices in the United States on prices corrected for exchange rates abroad:

$$P_{us} = \alpha + \beta(e_{xchangerate})(P_{elsewhere}) + \varepsilon.$$

Why would you want to know such a thing? If you were the chairman of the Federal Reserve Board and wanted to know whether your policy was controlling inflation, it would be more than a matter of idle curiosity to know whether foreign inflation rather than your bold manipulation of the federal funds rate was important. The null hypothesis of such a test is β = 1, that is, that foreign inflation causes American inflation, adjusted for the exchange rate, in exact proportion. A 1 percent inflation in the price level of the world causes, with a given exchange rate, is a 1 percent inflation in the United States. If the coefficient is lower—far above or below 1.0—the relation is not exactly described by purchasing power parity. (But beware: a strange relationship, β = −3.456, say, is still a relationship and could be substantively important for macroeconomic policy.)

In a well-known article of 1978 Irving Kravis and Richard Lipsey performed such a test. At conventional levels of *t* the Kravis-Lipsey equation "fails." Purchasing power failed the Whether test. But unfortunately that is the metaphysical not the scientific question. The scientific question— How close to exactly 1.00 does β have to be before one "accepts" purchasing power parity?—did not get answered. It is not the same question as "How high is the *t*-statistic for a null of β = 1.00?" The scientifically relevant question is a question of how big the parameter of interest is, not the Fisherian question of how probable the data are, given the null hypothesis, a purely sampling problem, constrained by the stability of the underlying distributions and so forth. You can ask how much oomph the foreign price has. Or you can ask how frequently the oomph is modified by "sampling error" on the strange assumption that a time series *is* a "sample" of some universe. The first question is clearly the relevant one. Yet Kravis and Lipsey, along with most of their colleagues, substituted the second, sampling question for the first, oomph question.

Being good economists, Kravis and Lipsey were evidently made a little uncomfortable by their own rhetoric. They admitted that "each analyst will have to decide in the light of his purposes whether the purchasing power relationships fall close enough to 1.00 to satisfy the theories" (1978, 214). That's right. In the next sentence, though, they again lost sight of the need for an explicit, substantive standard of How Big: "As a matter of general judgment we express our opinion that the results do not support the notion of a tightly integrated international price structure." They did not say in the text what the quantitative standard for their judgment was. They drifted away from oomph and toward the standard error. They substituted metaphysical judgments along the probability line, 0 to 1.0, for scientific judgments of how big is big in a regression coefficient.

In a footnote they reported the judgments of the economists Hendrik Houthakker, Gottfried Haberler, and Harry Johnson that a β within the range 0.8 to 1.2 would pass muster. And in fact most of their fitted equations fell within such a range. They nonetheless concluded, "We think it *unlikely* that the *high* degree of national and international arbitrage that many versions of the monetarist theory of [*sic:* they meant "the Mundell-Johnson monetary approach to"] the balance of payments contemplate is *typical* of the real world. This is not to deny that the price structures of the advanced industrial countries *are linked* together, but it is to suggest that the links are *loose* rather than *rigid*" (Kravis and Lipsey 1978, 243; italics supplied). Every italicized word involves a quantitative comparison along some scale of what constitutes unlikelihood or highness or typicality or looseness or rigidity. Yet nowhere in their article, and hardly ever in the tortured literature of purchasing power parity, was a standard proposed—other than the irrelevant one of R. A. Fisher's Rule of Two.

To be quite fair, Kravis and Lipsey were well above the average—some two or three standard deviations, we would judge—in their sensitivity to the need to have *some* standard. They returned to the issue repeatedly, if uneasily. But they rejected in one unpersuasive sentence the only substantive standard so far proposed in the literature—the so-called Genberg-Zecher criterion of whether *international* price integration is much less than integration within a single country. If Britain and Japan and the United States are as closely integrated in prices as are Belfast–London or Nagasaki–Hakadote or New York–Los Angeles then economically speaking there is no sense to treating the international price levels as independent.[1] If a country is treated as a geographical point in space, then so should the two countries with similar correlations of prices. That is a quantitative standard giving an economically relevant scale along which to judge How Big.

Kravis and Lipsey did distinguish between economic and statistical significance.[2] Even small differences between domestic and export prices, they noted, can make a big difference in the incentive to export: "This is a case in which statistical significance does not necessarily connote economic significance" (1978, 205). Precisely. If the world showed a β of 0.99999 but the econometrician had a gigantically large sample and produced therefore a standard error of .00000001, she would presumably *not* conclude that purchasing power parity had "failed."

Yet this is what the usual tests do. In another article of 1978 J. D. Richardson regressed Canadian prices on American prices corrected for the exchange rate and concluded, "It is notable that the 'law of one price' [another name for purchasing power parity] fails uniformly. The hypothesis of *perfect* commodity arbitrage is rejected with 95 percent confidence for every commodity group" (1978, 347; italics supplied). But why would it matter in an imperfect world if "perfect" arbitrage, β = 1.000000, say, were "rejected" by the irrelevant standard of sampling variability?

Jacob Frenkel is an enthusiast for purchasing power parity as such things go among economists. When late in his career he became president of the Bank of Israel he doubtless applied it to policy recommendations. But in 1978 Frenkel, like more than 70 percent of his colleagues publishing in the next decade in the *American Economic Review,* was bewitched by the ceremony of statistical significance. He wrote that "if the market is efficient and if the forward exchange rate is an unbiased forecast of the future spot exchange rate, the constant [in an equation of the spot rate today as a function of the future rate for today quoted yesterday] . . . should not differ *significantly* from unity" (1978, 175; italics supplied). Here he means "not *statistically* significantly different from 1.00," which is what he finds. In a footnote on the next page, though, he argues that "while these results [are statistically significant] . . . the 2–8 percent errors were *significant.*" Here he means "economically significant in some unspecified sense, perhaps as offering opportunities for profit."

Similarly Paul Krugman wrote in 1978 that "there are several ways in which we might try to evaluate purchasing power parity as a theory. We can ask how much it explains [i.e., by using the irrelevant and noneconomic criterion of fit]; we can ask how large the deviations from . . . parity are *in some absolute sense;* and we can ask if the deviations . . . are *in some sense* systematic" (1978, 405; italics supplied). The defensive phrases "in some absolute sense" and "in some sense" betrays his unease. There is no "absolute" sense in which a description is good or bad. France is a hexagon, roughly, for some purposes; for others not. The sense must be compared to a standard, and the standard must be economically relevant.

No economic hypothesis predicts that some β will be equal to 1.000000 "absolutely." Any scientific hypothesis is a matter being close enough. The decisions the scientist makes on what constitutes "closeness," as Abraham Wald said in 1939, "depend entirely on the special purposes of the investigator" (1939, 302). For purposes of making money on the foreign exchanges it might be necessary to have great exactitude, $\beta = 1.000 \pm .002$, say. But for the purpose of properly allowing for the international causes of American inflation it may be adequate to have a much rougher idea, $\beta = 0.90 \pm 0.20$, say. For such a purpose, in fact, if β—the influence of foreign prices on American prices—were almost anything nonzero it would matter for economic policy. As we said, a loose or strange relation is still a relationship. Gosset and Friedman knew what to do with a loose or strange relation. Fisherians, like most economists, don't.

9

What the Sizeless Stare Costs,
Economically Speaking

*The economic approach seems (if not rejected owing to aristocratic
or puritanic taboos) the only device apt to distinguish neatly what
is or is not contradictory in the logic of uncertainty.*

BRUNO DE FINETTI 1971, 486–87

You cannot "test" mechanically for nonzero along some scale that has no
dimension of substance and cost. How many molecules do you suppose
you share with William Shakespeare? We mean molecules in your body
that were once in his? Surprisingly, the correct answer, in view of the immense number of molecules in a human body and the operation of decay
and Brownian motion, is "quite a few." But for most questions—such as
"What is the chance I will be the next Shakespeare?"—the correct answer
is "negligible; *roughly* zero." Real scientific tests are always a matter of
how close to zero or how close to large or how close to some parameter
value, and the standard of how close must be a substantive one, inclusive
of tolerable loss.

IF THE EMPLOYMENT SUBSIDY IS GOOD
FOR WHITE WOMEN, THEN . . .

Testing economic hypotheses in no particular dimension yields a spaced-
out economics, and points to the wrong policies. Consider, for example,
an article in the *American Economic Review* in the 1980s that estimated
benefit-cost ratios in an Illinois experiment concerning unemployment insurance. We have mentioned it several times. In brief, the experiment paid

a cash bonus for giving an unemployed person a job (September 1987). In the so-called Employer Experiment the firms were given "a marginal wage-bill subsidy, or training subsidy, . . . [in order to] reduce the duration of insured unemployment" (517). In the control group the workers were paid the same subsidy, only this time the check was mailed directly to their homes. The main benefit of the training subsidy from the point of view of the state of Illinois was the reduction in unemployment benefits needed once the worker is back in employment. To this should be added, among other things, the benefit in the self-respect of the worker and the amount her labor adds to state economic output. But suppose we consider only the simplest cash accounting, as it appears the authors did. The "cost" of the experiment was the dollars of tax money spent on the subsidy. So a benefit-cost ratio of 4.29 means that the state saved $4.29 for each dollar it spent.

Here is how the authors interpreted their findings: "The fifth panel . . . shows that the overall benefit-cost ratio for the Employer Experiment is 4.29, *but it is not statistically different from zero*. The benefit-cost ratio for white women, . . . however, is 7.07, *and is statistically different from zero*. . . . The Employer Experiment *affected* only white women" (1987, 527; italics supplied). The 7.07 ratio "affects," they said, the 4.29 did not. This is a mistake. The best guess of the researchers was that the state got $4.29 for every dollar spent. The estimate was fuzzy, speaking of random sampling error alone. But that *does not mean it is to be taken as zero*. The program worked very well. By reporting that the 4.29 ratio was not "significant," and therefore supposing that it was in fact zero, and therefore not telling the policymakers that they should use it, the economists hurt the taxpayers of Illinois and immiserized the unemployed.

A fair question to ask of the Illinois experiment is *how noisy?* Just how weak was the signal-to-noise ratio, assuming that one thinks the measure is captured by the calculations of sampling error? The answer underscores the arbitrariness of Fisher's 5 percent ideology—the Type I error was about 12 percent ($p \leq .12$). That is to say, the 4.29 benefit-cost ratio was in the pilot study statistically significant at about the .12 level. In other words, the estimate was not all that noisy. A pretty strong signal for a very strong employment program. It was ignored.

By contrast, Joshua D. Angrist (three scores of Good in the 1990s) does well, in his "The Economic Returns to Schooling in the West Bank and Gaza Strip" (1995, 1065–1087), asking a question of oomph right from the outset. "Until 1972," Angrist writes, "there were no institutions

of higher education in these territories. . . . By 1986, there were 20 institutions granting post-high school degrees in the territories. As a consequence, in the early and mid 1980's, the labor market was flooded with
new college graduates. This article studies the impact" (1064). In a first
regression Angrist estimates the magnitude of wage premiums earned by
the influx of skilled workers on the distribution of wages in Israel.

> The first column of Table 2 shows that the daily wage premium for
> working in Israel fell from roughly *18 percent* in 1981 to zero in 1984.
> Beginning in 1986, the Israel wage premium rose *steeply*. By 1989, daily
> wages paid to Palestinians working in Israel were *37 percent higher* than
> local wages, nearly doubling the 1987 wage differential. The monthly
> wage premium for working in Israel *increased similarly*. These changes
> parallel the pattern of Palestinian absences from work and are consis-
> tent with movements along an *inelastic* demand curve for Palestinian
> labor. (1072, italics supplied)

No mention of statistical significance. The reader is told the magnitudes.
She knows the oomph. The inelastic demand curve, not the exact p-value,
is the object of policy relevance.

Yet even Angrist falls back into asterisk econometrics (his article is
Good, not Exemplary). On page 1079 he is testing alternative models and
emphasizes that "the alternative tests are not significantly different in five
out of nine comparisons ($p < 0.02$), but the joint test of coefficient equality for the alternative estimates of [θ_t] leads to rejection of the null hypothesis of equality" (1079). To which his better nature should say, "*Nu?*"

Solon Gets It

David Zimmerman, in his "Regression toward Mediocrity in Economic
Stature" (1992), and especially the well-named Gary Solon, in his "Intergenerational Income Mobility in the United States" (1992), set a rare standard for the field. Zimmerman revisits Galton's seminal study of "rever-
sion" in stature in fathers and sons, replacing "stature" with earnings
from labor. Solon asks a similar question, focusing on intergenerational income mobility. How much, Solon wonders, is a son's economic well-being
fated by that of his father? Line by line Solon asks the question "how
much?" and each time gives an answer. Previous estimates, observes
Solon, had put the father-son income correlation at about 0.2 (394). A
new estimate, a tightly fit correlation of 0.20000000001***, would say

nothing new of *economic* significance. And a well-fitting correlation with the "expected sign" would say nothing at all.

Solon's attempts at a new estimate, on pages 397–405, refer only once to statistical significance (1992, 404). Instead, he writes eighteen paragraphs on *economic* significance: why he believes the "intergenerational income correlation in the United States is [in fact] around 0.4" (403) and how the higher correlation changes the optimistic American stories about mobility—in which individual initiative, not your father's income, is supposed to matter. Notice the respect for the approximate nature of social statistics in his very phrasing of "around 0.4" instead of the 0.40768934 his computer undoubtedly spewed out. Solon's article, we would guess, is three or four standard deviations above the average of the *AER*. It changed our minds, which is one reasonable test for the quality of a scientific article read by scientists in the relevant field. Before reading the article one of us was very unwilling to believe that America was anything less than a land of opportunity. Real science changes ones mind. That's one way to see that the proliferation of unpersuasive significance tests is not real science.

"Minimum Wages and Employment: A Case Study of the Fast-Food Industry in New Jersey and Pennsylvania" by David Card and Alan B. Krueger (1994a), by contrast, is about average for cogency in statistical testing. It does not in *significance testing* change one's mind. The article typifies the standard errors of the *AER*. But it is well above the average in many other features of scientific seriousness, and for those reasons it did change minds, though we expect that those changes already leaned toward the expected ideology. We would like to meet the Chicago School economist who now believes, having studied Card and Krueger, that marginal productivity curves do not slope down. In any event, Card and Krueger most admirably and scientifically designed their own surveys, collected their own data, talked on the telephone with firms in their sample, and visited firms that did and did not respond to their survey, all of which is most unusual among economists. Their example seems in recent years to have raised scientific standards in economics. It matches the typical procedure in economic history, for example, or the best in empirical sociology and experimental physics.

But not for its use of statistical significance. Their sample was drawn to study prices, wages, output, and employment in the fast food industry in eastern Pennsylvania and western New Jersey before and after New Jersey raised its minimum wage above the national, and more to the point above the Pennsylvanian, levels. On pages 775–76 of the article (and pp. 30–33 in their book on the subject [1994b]), they report their crucial test of the conventional labor market model. The chief prediction of the conventional model is that full-time equivalent employment in New Jersey relative to Pennsylvania would *fall* following the increase in the New Jersey minimum wage. Specifically their null hypothesis says that the difference in difference is zero—that "change in employment in New Jersey" minus "change in employment in Pennsylvania" should equal zero if, as they suppose, the minimum wage does *not* have economic oomph. If they find substantively that the difference in difference is zero (other things equal), then by raising the minimum wage the progressive state of New Jersey gets the wage gains *without* loss of employment—a good thing for workers. Otherwise, New Jersey employment under the raised minimum wage will fall, perhaps by a lot—a bad thing for workers, as conventional and free-market opinion in economics would expect.

Yet Card and Krueger fail to test the null they claim. Instead they test two distinct nulls, "change in employment in New Jersey = zero" and (in a separate test) "change in employment in Pennsylvania = zero." In other words, they compute t-tests for each state, examining average full-time equivalent employment before and after the increase in the minimum wage. But they do not test the (only relevant) difference in difference null of zero. They report on page 776 a point estimate suggesting that employment in New Jersey *increased* by "0.6" of a worker per firm, from 20.4 to 21.0—rather than falling as enemies of the minimum wage would have expected. Then they report a second point estimate suggesting that employment in Pennsylvania fell by 2.1 workers per firm from 23.3 to 21.2. "Despite the increase in wages," they conclude from the estimates, "full-time equivalent employment *increased* in New Jersey relative to Pennsylvania. Whereas New Jersey stores were initially smaller, employment gains in New Jersey coupled with losses in Pennsylvania led to a small and statistically insignificant interstate difference in wave 2" (1994a, 776; italics in original).

Card and Krueger ran the wrong test (recall that testing the wrong null was less common in the *AER* during the 1980s [table 6.1, question 4]). They reject a null of zero change in employment in New Jersey, having found an average difference, estimated noisily at $t = 0.2$, of 0.6 workers per firm. They do not discuss the power of their tests, though the Pennsylvania sample is larger by a factor of five. They practice asterisk econometrics (with a "small and statistically insignificant interstate difference").

A confidence interval around the New Jersey estimate could easily lead to the opposite conclusion—that average employment fell. And yet they emphasize *acceptance* of their favored alternative with italics. Further attempts to measure with multiple regression analysis the size of the employment effect, the price effect, and the output effect, though technically improved, are not argued in terms of economic significance. That's the main point after all: how small is small? Card and Krueger miss their chance to say.

Douglas B. Bernheim and Alan Wantz do similar work in their article "A Tax-Based Test of the Dividend Signaling Hypothesis" (1995). They report that "the coefficients [in four regressions on their crucial variable, high-rated bonds] are all negative. . . . However, the estimated values of these coefficients are not statistically significant at conventional levels of confidence" (543). The basic problem with sign econometrics, and with the practice here, can be imagined with two price elasticities of demand for, say, insulin, both estimated tightly so far as sampling variability is considered, one at size -0.1 and the other at -4.0. Both are negative, and both would be treated as "success" in establishing that insulin use "responded to price." It's Samuelsonian economics, qualitative theorems. But the policy difference between the two estimates is of course enormous. Economically (and medically) speaking, for most imaginable purposes -0.1 is virtually zero. But when you are doing sign econometrics you ignore the *size* of the elasticity, or the *dollar effect* of the bond rating, and say instead, "the sign is what I expected."

BECKER STILL DOESN'T GET IT

The cost to scientific enlightenment of following the wrong decision rule is especially clear in "An Empirical Analysis of Cigarette Addiction" (1994), by Gary Becker, Michael Grossman, and Kevin Murphy, which we mentioned earlier. You can see that we do not want to be accused of making our lives easy by picking on the less eminent economic scientists: we have criticized Card, Krueger, Kravis, Lipsey, Krugman, Bernanke, and Blinder, eminent all, and now Grossman, Murphy, and especially Becker. Becker, we have mentioned, was a Nobel laureate, and Murphy was a 2005 winner of a MacArthur Foundation "genius" award. We realize, of course, that in 1994 Becker, the senior coauthor, was not himself performing the econometrics. But we suppose he played a role with Murphy and Grossman in the econometric interpretation.

Sign and asterisk econometrics decide nearly everything in the Becker article, but most irrelevantly the "existence" of addiction.

> Our estimation strategy is to begin with the myopic model. We then test the myopic model by testing whether future prices are significant predictors of current consumption as they would be in the rational-addictive model, but not under the myopic model. . . . According to the parameter estimates of the myopic model presented in Table 2, cigarette smoking is inversely related to current price and positively related to income. (Becker, Grossman, and Murphy, 1994, 403–4)

And then: "The highly significant effects of the smuggling variables (ldtax, sdimp, and sdexp) indicate the importance of interstate smuggling of cigarettes." It equates economic with statistical significance, as if addiction to cigarettes is valued in no currency.

But, as Kruskal put it, echoing Neyman and Pearson from 1933, "The adverb 'statistically' is often omitted, and this is unfortunate, since statistical significance of a sample bears no necessary relationship to possible subject-matter significance of whatever true departure from the null hypothesis might obtain" (1968a, 240). With N = about 1,400, Becker, Grossman, and Murphy can with high power reject a nearby alternative to the null—an alternative different, but *trivially* different, from the null. At high sample sizes, of course, all null hypotheses are rejected, by mathematical fact, as we have noted, without having to look at the data. No magic of instrumental variables is going to change that. Yet the authors conclude that "the positive and significant past-consumption coefficient is consistent with the hypothesis that cigarette smoking is an addictive behavior" (1994, 404). They are indulging in sign econometrics, with policy implications.

When sign econometrics meets asterisk econometrics the mystification redoubles.

> When the one-period lead of price is added to the 2SLS models in Table 2, its coefficient is negative and significant at all conventional levels. The absolute t ratio associated with the coefficient of this variable is 5.06 in model (i), 5.54 in model (ii), and 6.45 in model (iii). These results suggest that decisions about current consumption depend on future price. They are inconsistent with a myopic model of addiction, but consistent with a rational model of this behavior in which a reduction in expected future price raises expected future consumption, which in turn raises current consumption, while the tests soundly reject the myopic model. (Becker, Grossman, and Murphy 1994, 404)

Eventually they report (though they do not interpret) the estimated magnitudes of the price elasticities of demand for cigarettes. But their way of finding the elasticities is erroneous. Becker, Grossman, and Murphy set themselves a positivist standard of showing from the outside the "rationality" of addiction. They have not done so. They think they have shown a Whether. But Whether is not quantitative science. (They are, incidentally, inferring individual behavior from statewide data, which sociologists call the ecological fallacy.) Perhaps what they have shown is that statistics play multiple roles.

> There are some other roles that activities called "statistical" may, unfortunately, play. Two such misguided roles are (1) to sanctify or provide seals of approval (one hears, for example, of thesis advisors or journal editors who insist on certain formal statistical procedures, whether or not they are appropriate); (2) to impress, obfuscate, or mystify (for example, some social science research articles contain masses of undigested formulas [or tests of significance] that serve no purpose except that of indicating what a bright fellow the author is). (Kruskal 1968b, 209)

10

How Economics Stays That Way:
The Textbooks and the Referees

Small wonder that students have trouble [learning significance testing]. They may be trying to think.
W. EDWARDS DEMING 1975, 152

The proximate cause of the unhappy situation in economics is that almost all the teachers of econometrics claim that statistical significance is the same thing as scientific significance. The econometrician David Hendry, for example, is famous for saying "test, test, test," where the phrase means "Fisher, Fisher, Fisher," and most statistical textbooks in any field, from advanced theoretical statistics down to the merest cookbook, recommend the same (Hendry 1980).

A few get it right. Morris DeGroot, a Roosevelt University graduate (1952) and a distinguished statistician and teacher of several Nobel laureates in economics at Carnegie-Mellon University, wrote as follows in his exemplary textbook of 1975.

> It is extremely important . . . to distinguish between an observed value of U that is statistically significant and an actual value of the parameter. . . . In a given problem, the tail area corresponding to the observed value of U might be very small; and yet the actual value . . . might be so close to [the null] that, for practical purposes, the experimenter would not regard [it] as being [substantively] different from [the null]. (496).

DeGroot does not leave the matter as a throwaway point, a single sentence in an otherwise Fisherian tract, as so many of even the minority of statistics books that so much as mention the matter do. On the contrary, he goes on, "It is very likely that the *t*-test based on the sample of 20,000 will lead to a statistically significant value of *U*. . . . [The statistician] knows in advance that there is a high probability of rejecting [the null] even when the true value . . . differs only slightly from [the null]" (497).

But few econometrics textbooks make the distinction between statistical and economic significance. Even the best do not give equal emphasis to economic significance, to balance the scores, sometimes hundreds, of pages devoted to explaining Fisherian significance. In the texts widely used in the 1970s and 1980s, for example, when bad practice was becoming standard, such as Jan Kmenta's *Elements of Econometrics* (1971) or John Johnston's various editions of *Econometric Methods* (1963, 1972, 1984), there are no mentions of the distinction. Peter Kennedy, in his *A Guide to Econometrics* (1985), briefly mentions that a large sample always gives "significance." This is part of the point, but not nearly all of it, and in any case it is relegated to an endnote (62). He says nothing else on the matter.

CLIVE GRANGER ON NOT MENTIONING ECONOMIC SIGNIFICANCE

Arthur Goldberger gives the topic of "Statistical vs. Economic Significance" a page of his *A Course in Econometrics* (1991), quoting a little article by McCloskey in 1985. Goldberger's lone page has been flagged as unusual. The same Clive Granger reviewed four econometrics books in the March 1994 issue of the *Journal of Economic Literature* and wrote that "when the link is made [in Goldberger between economic science and the technical statistics] some important insights arise, as for example the section [well . . . the page] discussing 'statistical and economic significance,' *a topic not mentioned in the other books*" by R. Davidson and J. G. MacKinnon; W. H. Greene; and W. E. Griffiths, R. C. Hill, and G. G. Judge (Granger 1994, 118; italics supplied).

Not mentioned. *That* is the standard for education in econometrics and statistics at the advanced level. The three stout volumes of the *Handbook of Econometrics* (Griliches and Intriligator 1983–86) contain a lone mention of the point, unsurprisingly by Edward Leamer.[1] In the 732 pages of the *Handbook of Statistics* there is one sentence by Florens and Mouchart (Maddala, Rao, and Vinod 1993, 321). Aris Spanos has in his impressive *Probability Theory and Statistical Inference* tried to crack the Fisher monopoly on advanced econometrics, but even Spanos, a Hendry student, looks at the world with a sizeless stare (1999, 681–728). His history of hypothesis testing has in any case been ignored.

In the heyday of rational expectations macroeconomics, for example, its leading practitioners did not get the statistical point even approximately right (Lucas and Sargent 1981). No wonder. As in so many other fields, the econometrics of rational expectations has been entirely inconclusive. Many articles were produced, many *t*-tests performed, many careers smoothly advanced. No scientific findings yet. For example, in "Rational Expectations, the Real Rate of Interest, and the Natural Rate of Unemployment," even so fine an economic scientist as Thomas Sargent is thrilled to find that the "*F* statistic for regression (3) . . . is now 4.5, which exceeds the value of 2.503 [given to four significant digits, observe] . . . and so is even more significant statistically. *Thus, the test continues to point toward rejection of the natural rate hypothesis*" (Sargent 1981, 197; italics supplied). Uh huh.

One might argue that our point is so elementary and obvious that advanced books take it as given—though note that Sargent in 1981 didn't understand it (and neither, come to think of it, did in 1981 or 1964–68 his graduate school classmate McCloskey). Economy of style would dictate that the unqualified word *significance,* its exact meaning, economic or statistical, be supplied by the sophisticated reader. Under such a hypothesis the contemporary usage would be no more than a shorthand way to refer to an estimated coefficient.

In the elementary courses is the elementary point made? Takeshi Amemiya's advanced textbook in econometrics (1985), for instance, never distinguishes economic from statistical significance. He recommends that the student prepare for his highly theoretical book "at the level of Johnston, 1972." Examine Johnston, then, as Amemiya recommends, in Johnston's most comprehensive edition (1984). Johnston uses the term *economic significance* once only, rather late in the book, while discussing a technique commonly used outside of economics, without contrasting it with statistical significance, on which he has lavished by then hundreds of pages: "It is even more difficult to attach *economic* significance to the linear combinations arising in canonical correlation analysis than it is to principal components" (333; italics supplied).

At the outset, in an extended example of hypothesis testing spanning pages 17 to 43, Johnston goes about testing in the orthodox Fisherian way. In a toy example he tests the hypothesis that "sterner penalties" for dangerous driving in the United Kingdom would reduce road deaths, and concludes that "the computed value [of the *t*-statistic] is suggestive of a reduction, being significant at the 5 percent, but not at the one percent level" (Johnston 1984, 43).

What does this mean? Johnston suggests that at a more rigorous level—1 percent—you might not act on the result, *although acting on it would have saved about 100,000 lives in the* United Kingdom over the period 1947–57. Johnston has merged statistical and policy significance. Sterner penalties, according to his data, save lives. The rigorously 1 percent statistician, Johnston implies, would ignore this fact. By what warrant?

Johnston does recommend at the end of his book a sensible "Cairncross Test" (1984, 509–10). Sir Alec Cairncross (1911–98) was an economic historian and professor of economics of legendary learning and common sense, an economic consultant greatly admired in British and international circles. His doctoral dissertation was a pioneering piece of historical economics. The Cairncross criterion was: "Would Sir Alec be willing to take this model to Riyadh?" That is, would he use it for advising on real economic development in real places? But significance testing, Sir Alec told one of us, is not what he had in mind. He reported of his experience at His Majesty's Treasury after the war that his fellow economists, fresh from the first courses in econometrics, would give in the morning one significant estimate of the elasticity of demand for British exports and quite another significant one by evening (Cairncross 1992). Cairncross didn't believe either.

A tenacious defender of the prevailing method might argue that Johnston in turn had assumed that his readers got their common or Gosset sense from still more elementary courses and books. Johnston directs the reader who has difficulty with his first chapter to a "good elementary book" (1984, ix), mentioning Hoel's *Introduction to Mathematical Statistics* (1954), Mood's *Introduction to the Theory of Statistics* (1950), and Fraser's *Statistics: An Introduction* (1958). These are fine books. McCloskey first learned mathematical statistics from a later book co-authored by Mood, although she would not describe that one, or the others, quite as "elementary." The Mood and Graybill book gives a treatment of power functions, for example, which a modern economist would do well to read.

But none of the three books, or Mood and Graybill, makes a distinction between substantive and statistical significance. Hoel, for example, writes:

There are several words and phrases used in connection with testing hypotheses that should be brought to the attention of students. When a test of a hypothesis produces a sample value falling in the critical region of the test, the result is said to be *significant;* otherwise one says that the result is *not significant.* (1954, 176; italics in original)

R. A. Fisher Significance. That Is All

The old classic by W. Allen Wallis and Harry Roberts, which we have mentioned a couple of times, *Statistics: A New Approach,* first published in 1956, is an exception.

> It is essential not to confuse the statistical usage of "significance" with the everyday usage. In everyday usage, "significant" means "of practical importance," or simply "important." In statistical usage, "significant" means "signifying a characteristic of the population from which the sample is drawn," *regardless of whether the characteristic is important.* (1956, 385; italics supplied)

The point has been revived in some elementary statistics books, though not in most of them. In their leading book the statisticians David Freedman, Robert Pisani, and Roger Purves (1978) could not be plainer. In one of the numerous places in the first edition in which they make the point they write:

> This chapter . . . explains the limitations of significance tests. The first one is that "significance" is a technical word. A test can only deal with the question of whether a difference is real, or just a chance variation. *It is not designed to see whether the difference is important.* (487; italics supplied; compare 501, Appendix, 23)

The distinction is emphasized, as we've said, in the elementary books by Ronald J. Wonnacott and Thomas H. Wonnacott (1982, 160 [one of the brothers is an economist, the other is a statistician]) and admirably in David S. Moore and George P. McCabe (1993, 474). In econometrics Jeffrey Wooldridge (2000) is another standout, comparatively, we have noted, devoting about three pages to the matter. But three pages out of scores or hundreds? Is that the right proportion?

Some simple souls in other fields got it right. Economists are an arrogant lot and think of sociologists, psychologists, and educational researchers as beneath them in sophistication. But researchers in these fields have considered the difference between substantive and statistical significance.[2] Empirical sociology would be less easy for economists to sneer at if more economists realized that a good many sociologists grasped the elementary statistical point decades before even a handful of the economists did (Morrison and Henkel 1970). Psychologists have known about the difference for many decades, although, like economists, most of them continue to ignore it.

A Few Economists Have Protested

Of late the protest in and around economics has grown a little louder, but it is still scattered. The statisticians James Berger and Robert Wolpert, in 1984, though making a slightly different point (the Bayesian one that Jeffreys and Zellner emphasize), noted the large number of theoretical statisticians engaging in "discussions of important practical issues such as 'real world' versus 'statistical' significance": Schlaifer (1959); Pratt, Raiffa, and Schlaifer (1961 [1995]); Edwards, Lindman, and Savage (1963); I. J. Good (1981); Simonoff (2003); and the like. What we find bizarre, though, is that in the mainstream statistical literature this "important" point is hardly mentioned, and it is ignored by econometricians.

Among economists the roll of honor is likewise short relative to the thousands who have misused it. Consider F. Y. Edgeworth, Abraham Wald, Bruno de Finetti, J. M. Keynes (virtually), Ragnar Frisch, Oskar Lange, Leonard "Jimmie" Savage, Arnold Zellner, Arthur Goldberger, A. C. Darnell, Clive Granger, Edward Leamer, Milton Friedman, Robert Solow, Kenneth Arrow, Morris DeGroot, Howard Raiffa, Thomas Schelling, Zvi Griliches, Jack Hirshleifer, Glen Cain, Gordon Tullock, Lester Telser, Gary Solon, Joel Horowitz, Daniel Hamermesh, Jeffrey Wooldridge, Scott Gordon, Thomas Mayer, Erik Thorbecke, Nathan Berg, Allan Würtz, David Colander, Jan Magnus, and Hugo Keuzenkamp—all these are not dunces and they haven't minced words.[3] Recently, to pick one among the small, bright stream of revisions of standard practice that appear in our mailboxes, Clinton Greene (2003) has applied the argument to time-series econometrics, showing that tests of cointegration based on arbitrary levels of significance miss the economic point. The tests are neither necessary nor sufficient for scientific findings.

A famous econometrician told us recently that he didn't bother to teach The Point because his students at a leading graduate school were "too stupid" to do anything but the 5 percent routine. We find his response unreasonable and more than a little unethical. In 1986 the late Zvi Griliches made The Point but in a confused way illustrating the failure of the leaders of economics to take responsibility for their erroneous teachings: "Here and subsequently, all statements about statistical 'significance' should not be taken literally. . . . Tests of significance are used here as a metric for discussing the relative fit of different versions of the model. In each case, the actual magnitude of the estimated coefficients is of more interest" (Griliches 1986, 146). Notice his use of the comparison-of-models defense. Griliches had acted earlier as a commentator on a presentation by McCloskey, in 1984, at the American Economic Association annual meetings in Dallas. The experience appears to have made him sensitive to the charge that R. A. Fisher tests have a narrow and usually scientifically irrelevant meaning. But he never did explain the sense in which tests of significance could be used as "a metric for relative fit" and in particular why one would care about relative "fit" rather than "the actual magnitude of the estimated coefficients."

What is going on? We asked William Kruskal a couple of years before his death, "Why did significance testing get so badly mixed up, even in the hands of professional statisticians? Why did your devastating survey on 'significance' in the *International Encyclopedia of the Social Sciences* (1968a, with a version in Kruskal and Tanur 1968 [1978]) have no effect?" "Well," replied Kruskal, smiling sadly, "I guess it's a cheap way to get marketable results."

So it would seem. Finding statistical significance is simple, and publishing statistically significant coefficients survives at least that market test. Cheap *t*-tests, becoming steadily cheaper with falling computational costs, have in equilibrium a marginal scientific product equal to their cost. Entry ensures it. Edgeworth said so at the dawn. He corrected Jevons, who had concluded that a "3 or 4 per cent" difference in the volume of commercial bills is not economically important: "[B]ut for the purpose of science, the discovery of a difference in condition, a difference of 3 per cent and much less may well be important" (Edgeworth 1885, 208). It is easy to see why: a statistically *insignificant* coefficient in a financial model, for example, may nonetheless give its discoverer an edge in making a fortune. And a statistically *significant* coefficient in the same model may be offset in its exploitation by transaction costs.

The Miseducation of an Econometrician

In economics the problem does not originate in the late 1930s with the Cowles Commission. It originates in the writings of Harold Hotelling ten years earlier.[4] But at Cowles the problem certainly worsened. In 1936 a Second Annual Research Conference on Economics and Statistics was held, a Cowles conference, in Colorado Springs. In attendance were many luminaries such as Irving Fisher, Corrado Gini, and Ragnar Frisch. Harold Hotelling was a large presence, but a still larger one was Hotelling's mentor on matters statistical, Ronald A. Fisher, who gave a paper on "The Significance of Regression Coefficients" (Fisher 1936).

The significance of Cowles econometrics, and the influence of the Fisher-Hotelling relationship, is exemplified by the rhetorical history of Trygve Haavelmo's classic, "The Probability Approach in Econometrics" (1944), for which Haavelmo was awarded the Nobel Prize in 1989. Mary Morgan writes, "Haavelmo's work marks the shift from the traditional role of econometrics in *measuring* the parameters of a given theory [which is the role of such techniques in Gosset's guinnessometrics and in much of chemistry and physics] to a concern with *testing* those theories" (1990, 257; italics supplied). Haavelmo's article was crucial for bringing probability into the foundation of econometric thinking. Yet Morgan gives no sign in her otherwise comprehensive history of the coming of modern econometrics, Jevons to Haavelmo, that she grasps the absurdities of a Fisherian test criterion that was quickly adopted by the probabilistic econometricians.

Haavelmo himself certainly didn't grasp it. He initially hoped to develop a systematic approach to scientific explanation using—get this—*personal* probability (1958, 357). Later, in his classic article of 1944, he believed he was adopting "Neyman-Pearson" procedures. But he appears not to have realized that Neyman and Pearson were both anti-Bayesian and anti-Fisherian.[5] And so Haavelmo, Tjalling Koopmans, and the pioneer of most things econometric, Jan Tinbergen (who was awarded the Nobel Prize in 1969), adopted Fisher-significance as though Neyman and Pearson, or before them de Finetti, or before him Gosset, had never written (Koopmans 1937, in Hendry and Morgan, eds., 1995). That's what we mean when we say with Morgan that Haavelmo "marks a shift": though clearly he had intentions to work in Bayesian-personal-probabilistic and Neyman-Pearson frameworks, the fact is that he and the luminaries at the Colorado meetings adopted Fisher's method as the one and only.

Tinbergen, like Haavelmo, is a paradox, too. Tinbergen, Morgan notes, employed in 1939 the "classical" (her revealing name for the Fisherian) tests, though also an amazing fifteen other tests, including the comparing of simulations in the style of physics and engineering (Morgan 1990, 113). Tinbergen, like many of the early econometricians, began life as a physicist, as did his student Tjalling Koopmans (Nobel Prize, 1975), the great propagandist for the division of the empire of economics into econometric and mathematical-economics provinces. The fathers of econometrics—Frisch excepted—ignored their better training (Frisch [1934]). James Buchanan and Aris Spanos have drawn our attention to a debate in 1949 between a deeply Fisherian Koopmans and the Virginia economist Rutledge Vining during the controversy over the introduction of Cowles Commission methods as against the descriptive empiricism of the National Bureau of Economic Research (Vining 1949). Vining attacked Koopmans's proposed "strait jacket on economic research"—which subsequently the profession donned with exultation—and then quoted George Udny Yule, a Pearson protege, speaking in the early 1940s: "[T]here has been a completely lopsided—almost a malignant—growth of sampling theory. . . . Caution, common sense and patience . . . are quite likely to keep [the experimenter] more free from error . . . than the man of little caution and common sense who guides himself by a mechanical application of sampling rules. He will be more likely to remember that there are sources of error more important than fluctuations of sampling."[6] William Sealy Gosset couldn't have put it better.

Lawrence Klein (Nobel Prize, 1980) reinforced the Fisherian techniques for economics in the Samuelsonian age of economics, 1947 to the present. Klein was the first PhD student of Paul Samuelson (Nobel Prize, 1970) at MIT. Samuelson suggested to Klein that he apply Tinbergen's methods to American data. In his very first published scientific paper, in 1943, Klein uses words that were to become formulaic for those who followed him: "The role of Y in the regression is not statistically significant. The ratio of the regression coefficient to its standard error is only 1.812. This low value of the ratio means that we *cannot* reject the hypothesis that the true value of the regression coefficient is zero."[7]

That Samuelson had a hand in bringing *t*-testing to American economics is perhaps significant. At the time Samuelson was busy claiming that economic theory could be qualitative, a fatal turn in economics made official by his modestly entitled *Foundations of Economic Analysis* (1947) and reinforced by the existence-theorem techniques of his influential brother-in-law, Kenneth Arrow. In other words Samuelson, through Klein, brought the qualitative turn to empirical work as well. It does not matter, said Klein and Samuelson (Arrow on this statistical point in fact demurring), how big the effect of coefficient on Y is. What matters is whether its effect is there at all, discernible, significantly different from zero. What matters, they said, is Whether, not How Much.

Deming Tried to Save Management

Around the time that Hotelling, Haavelmo, Tinbergen, Koopmans, and Klein were taking econometrics down the Fisherian road, W. Edwards Deming was trying to take business management into the Sealy Gosset promised land (Deming 1938, 1961, 1982; see also Andrea Gabor's lucid intellectual biography [1992]).

In 1938 Deming published a little book on statistical methods, *Statistical Adjustment of Data*. Today people know Deming as the father of quality control, a role that suits him only if one thinks of Walter Shewhart (1929, 1931) as the grandfather. Shewhart got there first, at Bell Telephone, before he taught an already mature but curious Deming everything he could. Deming was easy to teach. Like Shewhart, he had been trained as a physicist and engineer, and by the late 1930s he was already an applied statistician of the first rank. When Ronald Fisher himself met Deming, the Englishman was simultaneously seduced and repelled by Deming's no-nonsense, Iowa farm boy style. In the 1938 book Deming warned business managers in no uncertain terms, as we have noted, that "Statistical 'significance' is by itself not a rational plan for action" (30). Like Gosset, whom he knew and admired, Deming saw that size matters in inductive decisions all the way down. And, also like Gosset, he saw that the scientist's job is to minimize the whole error—not merely sampling error from one experiment (29–31).

But in America and Europe, operation researchers and business planning teams would not listen. They were entranced by the easy gains imagined in Haavelmo-Kleinometrics. Events forced Deming to change his strategy (not his mind). Rhetorically speaking he became more waspish—imitating the forceful, authoritarian rhetoric of Ronald Fisher—to better steer management away from it. This he tried to do for the next fifty years. Only in Japan did he succeed.

His attempt late in life to bring the ideas back to his own country in *Out of the Crisis* was designed as a must read for the more statistically sophisticated American managers of its generation. It was an *In Search of Excellence* for techies. Published in 1982 and marketed heavily in 1986, *Out of the Crisis* was one of Deming's last attempts to "transform," as he puts it, the "Western style of management," to "halt the decline of Western industry" (18). The book offers "14 points" for "the removal of the deadly diseases and obstacles" to long-term profits and productivity.

Deming in 1982 believed that a widespread, Western industrial failure to understand "variation" was in fact the number-one "disease" or "obstacle" to industrial success. And yet to analyze the variation he did not accede to the use of tests of statistical significance, Fisherian or otherwise. On page 20 he says clearly (crediting Lloyd S. Nelson, the director of statistical methods for the Nashua Corporation, for helping him appreciate the point): "The central problem of management in all its aspects, including planning, procurement, manufacturing, research, sales, personnel, accounting, and law, is to understand better the meaning of variation, and to extract the information contained in variation." That is, watch closely when something is above or below the routine and jump on it.

Trained in statistics at the dawn of the Fisherian revolution, one would think that Deming, a longtime employee of the U.S. Department of Agriculture who once admired Fisher enough to travel to London to meet him and then spent a year with him in London on sabbatical, would merely follow the herd. He never did.

> There are many other books on so-called quality control [Deming wrote]. Each book has something good in it, and nearly every author is a friend and colleague of mine. Most of the books nevertheless contain bear traps, such as reject limits, . . . areas under the normal curve, acceptance sampling. . . . The student should also avoid passages in books that treat confidence intervals and tests of significance, as *such calculations have no application in analytic problems in science and industry.* (Deming 1982, 369; italics supplied)

Deming did not believe most firms in industry to be in a situation described by and required for the proper use of tests on experimental data (in this one sense he and Fisher agreed). Like Harold Jeffreys before him (1931, 1939a), Deming came to believe that management should not predict with the aid of classical tests of significance, Fisher's analysis of variance, and the like. Managers need instead "degrees of belief" (1982, 132). He saw a need to allocate the degrees of belief among hypotheses, given the data, rather than, as the Fisherians do, calculating the probability of the data given the hypothesis ("maximum likelihood" without a prior). He saw in quality control, in other words, what we have seen in economics, epidemiology, and many other fields: the fallacy of the transposed conditional married to the sizeless stare of statistical significance. He observed how "statisticians and management . . . misguide each other and keep the vicious cycle [of calculating and reporting t-tests] going" (133). Testimation closed their minds to the real variations that actually mattered.

Deming's earlier book, *Sample Design in Business Research* (1961), shows him as a longtime enemy of Fisher's null-hypothesis testing ritual. Published while he was a professor of statistics in the business school at New York University, the message of *Sample Design* to "statisticians themselves" and "executive[s] in business" (v) is plain.

> The standard error of a result does not measure the usefulness thereof. The standard error, however helpful in the use of data from samples, only gives us a measure of the variation between repeated samples. . . . It does not mean that the persistent components of the non-sampling errors are small. It is important, for such reasons, I believe, not to focus attention on the standard error alone. . . . In my own practice, I steadfastly refuse to compute or to discuss the interpretation of the standard error when large operational non-sampling errors are obviously present. (1961, 55–56)

Compare Gosset's "real" error. And in case the reader was lost in a fog of Fisher's sampling-only procedures, and therefore not paying attention, Deming made his point from the opposite direction, too, emphasizing the primacy of oomph, the "usefulness" criterion, even at the risk of allowing a wider Type I error.

> It is possible for a result to be useful and still to possess a wide standard error. A result obtained by definitions and techniques that have been drawn up with care, and carried out by excellent interviewing and supervision may have a wide standard error because the sample was small; yet such a result might well be preferable to one obtained with a bigger sample, with a smaller standard error, but whose definitions, techniques, and interviewing were out of line with the best practice and knowledge of the subject matter. (56–57)

Of course. In the preface to *Sample Design*, he wrote: "Statistical theory shows how mathematics, judgment, and substantive knowledge work together to the best advantage" (v). Or should.

Despite Deming, American Management Science Went *into* the Crisis

The main journals of management and organization theory today take the standard error as decisive. One of the leading journals recently published an article purporting to "test the Deming model" of management.

The article reveals little grasp of Deming's disdain for Fisher's error. Thomas J. Douglass and Lawrence D. Fredendall's "Evaluating the Deming Management Model," published in *Decision Sciences* (2004), "tests" the relevance of Deming's model to the service sector of the economy— hospitals in particular. Their figure 2, which "overlays *significant* paths onto the conceptual Deming management model" is off the shelf testimation (408–9; italics supplied). "The study," they explain, "was conducted within the General Medical Hospitals (SIC 8062) industry. [Total quality management] has been recommended to the members of this industry as a strategy that will assist them in dealing with their turbulent environment. . . . Thus, this context was expected to provide an excellent platform on which to test the subject model" (405).

The authors establish a nonexperimental setting for what they allege to be a quantitative, experimental test. They have no controls, whether experimental or observational, for determining the relative economic impact of the Deming model. Therefore they have no basis for judging How Much. That is, they do not know how much the hospitals improved. They do not watch for variation and jump on it. They depend throughout on a merely qualitative meaning of significance, exactly what Deming spent his career attacking. A table on page 409 features "a change in chi-square"—celebrating the increase in chi-square, in amounts above 5 percent significance, achieved as variables were added to regressions: "Top management team involvement, our measure of visionary leadership, was significantly related to both quality philosophy ($t = 10.80, p < .001$) and supplier involvement ($t = 7.59, p < .001$). Therefore, Hypothesis 1 is supported. . . . In our study, 'the authors continue' we did not find significant relationships between continuous improvement and financial performance or customer satisfaction. With respect to the. . . audit score ($t = 1.79, p < .10$), marginal significance was found. Therefore, Hypothesis 7 was generally not supported" (407–8). Deming himself asked of any service or product how in the eyes of the *user* it could be improved. No matter.

The authors prepared a table 3—results from the structural model— allegedly supporting their claims. But it is untranslatable as to oomph. It contains chiefly asterisks: ***, denoting $p < .001$; ** and * denoting $p < .01$ and $.05$, attached to numbers that are nowhere defined by size, units, averages, standard deviations, or meaning for profit or decision. In other words, in a major journal of the academic study of business the "significant" numbers are assigned no meaning except their Fisherian significance. The Deming model was not tested. It is an anti-Deming testimation of Deming.

If You Can't Go β, Go Scientific—with a Small s

We are sometimes told that "You're rehashing issues decided in the 1950s" or "But the hot *new* issue is [such and such instrumental variable technique]" or "I have a metaphysical argument about why a universe should be viewed as a sample." When we are able to get such people in a hurry to slow down and listen to what we are saying, which is not often, we discover that in fact they do *not* grasp our main point and that their own practice shows that they do not.

It is dangerous, for example, to mention Bayes in this connection because the reflexive reply of most econometrically minded folk is to say "1950s" or "I don't know my prior" and have done with it. Our main point is not Bayesian—although we honor the Bayesians such as Leamer and Zellner, who have made similar, and also some different, criticisms of econometric practice, and we do confess a prejudice. But our point about the sizeless stare (as against testimation) has nothing to do with Bayes's rule. It applies to the most virginal classical regressions.

Our experience is that in the rare cases when people *do* suddenly grasp our point—that fit and importance are not the same—they are appalled. McCloskey's colleague at the University of Illinois at Chicago, Lawrence Officer, describes himself this way, for example. They realize with a jolt that most of what has been done in research on the economy since the beginning of econometrics, including their own work, needs to be redone. The wrong variables have been included, for example, which is to say that errors in specification have vitiated the conclusions. Mistaken policies have been recommended. Science has been turned off the track.

We believe we have shown from the *American Economic Review* over the two last decades what economic scientists from Edgeworth to Goldberger have been saying: science is about magnitudes. Seldom is the magnitude of the sampling error the chief scientific issue. A reader sympathetic with the established view might reply that it's not the size that counts; it's what you do with it. But that, too, is mistaken. As Friedman's alloy regression and hundreds of other statistical experiments reveal, what matters is size *and* what you do with it. Scientific judgment, like any judgment, is about loss functions—what R. A. Fisher was most persistent in denying.

What should economists do? They should act more like the Gary Solons and the Claudia Goldins. They should be economic scientists, not calculators recording 5 percent levels of significance. In his acceptance speech for the Nobel Prize, Robert Solow put it this way.

> [Economists] should try very hard to be scientific with a small s. By that I mean only that we should think logically and respect fact. . . . Now I want to say something about fact. The austere view is that "facts" are just time series of prices and quantities. The rest is all hypothesis testing. I have seen a lot of those tests. They are almost never convincing, primarily because one senses that they have very low power against lots of alternatives. There are too many ways to explain a bunch of time series. And sure enough, the next journal will contain an article containing slightly different functional forms, slightly different models. My hunch is that we can make progress only by enlarging the class of eligible facts to include, say, the opinions and casual generalizations of experts and market participants, attitudinal surveys, institutional regularities, even our own judgments of plausibility. My preferred image is the vacuum cleaner, not the microscope. (1988 [1997], 203–4)

Solow recommends that economists "try very hard to be scientific with a small s"; the authors we have surveyed in the *AER*, by contrast, are trying very hard to be scientific with a small *t* (or a large one, if that's the way the null is set up).

As Solow says, it's almost never persuasive. In a way the lack of persuasion by *t* and *F* is encouraging. Despite the surface rhetoric of Fisherian tests that would imply a change of mind, no one, as we said, changes her mind. New data changes it. New theories. New authorities. New metaphors. New stories. New testimony. New experimental design. But never so far as we have been able to ascertain new *t*-statistics.

It has become a common challenge to put to the proud econometricians, gobbling up more and more of the graduate curriculum in economics and driving the students farther and farther away from confrontation of economic thinking with economic facts, what advance in economic science since the war has turned on a statistical test of significance? The answer appears to be: none. Implicitly, then, scientific practice in economics is agreeing with us: Fisherian tests are almost useless. Scientifically they have little point.

So what? This: economists might have avoided the faith of the 5 percenters—and evolved a quantitative science—had they listened to members of their own tribe such as Arrow and Zellner. In economics the quantitative revolution will not be complete until what counts as a critical value in its center is a demonstrated *economic* significance and not a so-called statistical significance.

Ziliak was advised to remove our 1996 *Journal of Economic Literature* article from his curriculum vitae while job hunting—it wasn't "serious" research. "Stay quiet, boy," was the message, "and follow R. A. Fisher." He was given similar advice in January 2004, this time as an associate professor. He was driving with other economists in an airport shuttle bus to the annual meetings of the American Economic Association, where he and McCloskey would address a very large assembly of economists. "It's not a very popular idea," a noted economist, seated nearby, remarked. "You're going to get slaughtered. *But,*" he snorted, "you can always blame it on McCloskey!" In the event we were nothing like "slaughtered." The session was a great success—although there is still no change in the behavior of significance testers.

McCloskey in the late 1990s served fleetingly on the editorial board of the *AER*. Each time she saw in a submission that the emperor had no clothes of oomph she said so. The trouble that McCloskey had with the routine of statistical testing in economics did not delight the editors. After a while she and they decided amicably to part. The *AER* continues to print articles dominated by meaningless tests of significance.

The situation is strange. Economic scientists—for example, those who submit articles to the *AER* or edit or referee it or some other journal or serve on hiring and grant-making committees—routinely violate elementary standards of statistical cogency. And yet it is the messengers who are to be taken out and shot. We have seen this strange sociology in other fields, such as in sociology itself.

This should cease. The economics profession should set meaningful standards of economic significance. If the *AER* or any one of a handful of other leading journals were to test articles for cogency and refuse to publish articles that used fit irrelevantly as a standard of oomph, economics would in a few years be transformed into a field with muscular empirical standards. True, bad practices are amazingly robust in science, so perhaps we are being optimistic. At present (we can say this until someone else starts claiming that *in the 2000s* practice in economics has improved) we have shown that economics has no scientifically relevant standards for its statistical work.

Ask: "Is the article mainly about showing and measuring *economic* significance?" If not, the editor and referees should reject—reject the article, that is. It will not reach correct scientific results. Its findings will be biased by misspecification and mistaken as to oomph. Requiring referees to complete our nineteen-item questionnaire would probably go against the libertarian grain of the field. A short form would do: "Does the article focus on the *size* of the economic effect it is trying to measure or does it instead recur to irrelevant tests of fit and a coefficient's *statistical* significance?" To do otherwise—continuing to decorate our articles with stars and *t*'s and standard errors while failing to interpret size—is to discard our best unbiased estimators and to renege on the promise of modern econometrics: measurement with theory. No size, we should say, no significance.

The Not-Boring Rise of
Significance in Psychology

The earth is round (p < *.05*).
JACOB COHEN 1994, 997

Economists, we have noted repeatedly, are not the only scientists to fall short of real significance. Psychologists have done so for many decades now. An addiction to transforms of categorical data, a dependence on absolute criteria of Type I error, and a fetish for asterisk psychometrics have been bad for psychology, as similar mistakes have been for economics.

Since Edwin G. Boring warned in 1919 against mixing up statistical and substantive significance the quantitative psychologists have been told by their own people again and again about the sizeless stare. Still they yawn—such a *boring* point, ha, ha. Since 1962, when Jacob Cohen published his blistering survey of statistical power in the field, psychologists have been shown in more than thirty additional studies that most of their estimates lack it (Rossi 1991). Between snores, few psychologists cared.

"SIGNIFICANCE IS LOW ON MY ORDERING"

Norman Bradburn, a psychologist and past-president of the National Opinion Research Center, a member and former chair of the Committee on National Statistics of the National Academy of Sciences, told a story about *p*-values in psychology at the memorial service for William Kruskal. Bradburn spoke of Kruskal's gentle demeanor. But "sometimes his irritation at some persistent misuse of statistics would boil over, . . . as with the author of an article that used *p*-values to assess the importance of differences"

(2005, 3). Bradburn himself was the author in question. "I'm sorry," wrote Kruskal, "that this ubiquitous practice received the accolade of use by you and your distinguished coauthors. I am thinking these days about the many senses in which relative importance gets considered. Of these senses, some seem reasonable and others not so. Statistical significance is low on my ordering. Do forgive my bluntness" (3).

Despite impressive attempts by such insiders to effect editorial and other institutional change—impressive at any rate by the standards of an economics burdened with cynicism in its worldview—educators and psychologists have produced pseudo-significant results in volume. Kruskal, though as we say a past-president of the American Statistical Association and a consummate insider, was too gentle to stop it. "Do forgive my bluntness" was as forceful as he got. Neither could Paul Meehl, though a famous academic psychologist, stop or even much slow down the beat of the 5 percenters. Meehl, by the way, was also a clinical psychologist and *was* able to help the difficult Saul Bellow—which astonished Bellow himself.[1] The persuasive Meehl became Bellow's model for Dr. Edvig in Bellow's novel *Herzog*: Edvig was "calm Protestant Nordic Anglo-Celtic." But changing the psychology of significance testing seems in psychology too much even for a calm Protestant Nordic Anglo-Celtic.

The Melton Manual

The history of the *Publication Manual of the American Psychological Association* exhibits the depth of the problem. The *Manual* sets the editorial standards for over a thousand journals in psychology, education, and related disciplines, including forensics, social work, and parts of psychiatry. Its history gives a half century of evidence that reform of statistical practice won't succeed if attempted by one science alone. It's embedded like a tax code in the bureaucracy of science. The failure of the Kruskals, Cohens, Meehls, and others contradicts our own optimistic hope that a change of editorial practices in the *American Economic Review* or the *Journal of Political Economy* would do the trick in economics. In psychology a large number of useful-sounding manifestos and rewritten editorial policies have not built a rhetoric or culture of size mattering.

In the 1952 first edition of the *Manual* the thinking was thoroughly pro-Fisher and anti-Gosset, obsessed with significance: "Extensive tables of non-significant results are seldom required," it says. "For example, if only 2 of 20 correlations are significantly different from zero, the two significant correlations may be mentioned in the text, and the rest dismissed with a few words."[2] The *Manual* was conveying what Fisher and Hotelling and others, such as Klein in economics and A. W. Melton in psychology, were preaching at the time. In the second edition—twenty years on—the obsession became compulsion.

> Caution: Do not infer trends from data that fail by a small margin to meet the usual levels of significance. Such results are best interpreted as caused by chance and are best reported as such. Treat the result section like an income tax return. Take what's coming to you, but no more. (*APA Manual* 1974, 19, quoted in Gigerenzer 2004, 589)

Recent editions of the *Manual*—as both critics and defenders of the establishment observe—do at last recommend that the authors report "effect size."[3] But as Gerd Gigerenzer, a leading student of such matters, observes, the fifth edition of the *Manual* retained also the magical incantations of $p < .05$ and $p < .01$. Bruce Thompson, a voice for oomph in education, psychology, and medicine, commends the fifth edition for suggesting that confidence intervals are the "best reporting strategy."[4] Yet, as Thompson and Gigerenzer and Fiona Fidler's team of researchers have noted, in Gigerenzer's words, "The [fifth] manual offers no explanation as to why both [confidence intervals for effect size and asterisk-superscripted p-values] are necessary . . . and what they mean" (2004, 594). The *Manual* offers no explanation for the significance rituals—no justification, just a rule of washing one's hands of the matter if $p < .05$ or $t > 2.0$.

In psychology and related fields the reforms of the 1990s were nice sounding but in practice ineffectual. The 2001 edition of the *Manual* appears to reflect pressure exerted by editors and scientists intent on keeping their machine for article-producing well oiled. Some twenty-three journals in psychology and education now warn readers and authors against the sizeless stare.[5] That is about 2 percent of the journals. The other 98 percent are sizeless. Despite the oomph-admiring language in recent editions of the *Manual*, published practice in the psychological fields is no better than in economics.

In 1950 A. W. Melton assumed the editorship of the trend-setting *Journal of Experimental Psychology*. In 1962 Melton described what had been his policy for accepting manuscripts at the journal (1962, 553–57). An article was unlikely to be published in his journal, Melton said, if it did not provide a test of significance and in particular if it did not show statistically significant results of the Fisher type. Significance at the 5 percent level was "barely acceptable"; significance at the 1 percent or "better" level was considered "highly acceptable" and definitely worthy of publication (544). Melton justified the rule by claiming that it assured that "the results of the experiment would be repeatable under the conditions described." Uh huh. The statisticians Freedman, Pisani, and Purves have observed sarcastically that "many statisticians would advise Melton that there is a better way to make sure results are repeatable: namely, to insist that important experiments be replicated."[6] The 5 percent/1 percent statistician ruled, and the scientific standard of replication fell away in psychology as it had in economics. Gigerenzer et al. (1989) note that after Melton's editorship it became virtually impossible to publish articles on empirical psychology in any subfield without "highly" statistically significant results. Some parts of psychology were spared: literary and humanistic psychology, for example. But we do not regard this as good news. The quantitative parts of a science should not be notable mainly for their lack of common sense.

In a penetrating article of 1959, "Publication Decisions and Their Possible Effects on Inferences Drawn from Tests of Significance—or Vice Versa," the psychologist Thomas D. Sterling surveyed 362 articles published in four leading journals of psychology: *Experimental Psychology* (Melton's journal), *Comparative and Physical Psychology, Clinical Psychology*, and *Social Psychology* (Sterling 1959). Table 11.1 shows his results. Everyone knows—but no one corrects their significance levels for it—that "significant" results are the only ones that see the printed page. The fact undermines the claim of significance since the so-called random sample is selected out of the numerous samples collected exactly for statistical significance. People in various statistical sciences in the 1950s and

TABLE 11.1. Outcomes of Tests of Significance in Four Psychology Research Journals, 1955–56

All Issues From	Number of Research Articles	Number Using Significance Tests	Number That Reject at ≤.05
Experimental Psychology (1955)	124	106	105
Comparative and Physical Psychology (1956)	118	94	91
Clinical Psychology (1955)	81	62	59
Social Psychology (1955)	39	32	31
Total	362	294	286

Source: Sterling 1959.

1960s, complained about publication bias, as the Fisherian machinery took hold and academic publishing expanded (in economics, for example; see Arrow 1959; and Tullock 1959). "The problem simply," Sterling explained, "is that a Type I error (rejecting the null hypothesis when it is true) has a fair opportunity to end up in print when the correct decision is the acceptance of H_0. . . . The risk stated by the author cannot be accepted at its face value once the author's conclusions appear in print" (1959, 34).

Sterling was therefore not surprised when he found that only 8 of 294 articles published in the journals and using a test of significance failed to reject the null. Nearly 80 percent of the articles relied on significance tests of the Fisherian type to make a decision (286 of 362 published articles). And, though Sterling does not say so, every article using a test of significance—that is, those 80 percent of all the articles—employed Fisher's 5 percent philosophy exclusively. (Melton's stricter rule of 1 percent was adopted by some of the journals.) The result "shows that for psychological journals a policy exists under which the vast majority of published articles satisfy a minimum criterion of significance" (Sterling 1959, 31).

Sterling observed further that despite a rhetoric of validation through replication of experiments—to which Gosset gave much of his scientific life, by the way, quite unlike Fisher, who preferred to do more elaborate statistical calculations on existing data—not one of the 362 research articles was a replication of previously published research.[7] From his data Sterling derived two propositions.

> A1: Experimental results will be printed with a greater probability if the relevant test of significance rejects H_0 for the major hypothesis with $Pr(E| H_0) \leq .05$ than if they fail to reject at that level.
> A2: The probability that an experimental design will be replicated becomes very small once such an experiment appears in print. (1959, 33)

He understated. Nearly certainly an experimental result that "fails to reject" will not be printed, and by A. W. Melton with probability 1.0. And why actually replicate when the logic of Fisherian procedures gives you a virtual replication without the bother and expense? Why not go ahead and use the alloys F1 and F2 in airplanes? After all, $p < .05$.

"A picture emerges," wrote Sterling with gentle irony, "for which the number of possible replications of a test between experimental variates is related inversely to the actual magnitude of the differences between their effects. The smaller the difference the larger may be the likelihood of repetition" (1959, 33). Sterling concluded that "when a fixed level of significance is used as a critical criterion for selecting reports for dissemination in professional journals it may result in embarrassing and unanticipated results" (31). In a recent study similar to Sterling's, Hubbard and Ryan (2000) found that in twelve APA-affiliated journals between 1955 and 1959 fully 86 percent of all empirical articles published had employed the 5 percent accept/reject ritual.[8] Educational psychology and other subfields of education had meantime taken the same turn.[9] They continue therefore to yield embarrassing and unanticipated—in plain words, wrong—results.

Some Psychologists Tried to Ban the Test

Joined by a few academic students of education, some psychologists, alarmed by the oil slick of the standard error, tried to ban it outright. Startlingly, the American Psychological Association arranged in the 1990s symposia to discuss the banishment of statistical significance testing from psychology journals. In 1996 an APA Task Force on Statistical Inference was appointed to investigate the matter. In 1997 *Psychological Science* published the proceedings of the first symposium. Some of the main critics of statistical significance, such as Jacob Cohen, Robert Rosenthal, Harold Wainer, and Bruce Thompson, served on its twelve-member jury. Such a selection would be highly unlikely in economics, where such committees become sites for exercise of power in aid of established ideas. A similar committee of the American Economic Association formed to investigate the over-formalization of graduate education in economics (e.g., the theoretical econometrics without training in other means of investigating economic phenomena) was torpedoed by some of the barons appointed to it. The task force of psychologists, in contrast, was not a whitewash.

It speaks well for the intellectual seriousness of psychology. McCloskey's colleague in psychology at the University of Illinois at Chicago, Chris Fraley, gives a detailed graduate course on null-hypothesis significance testing that would be very hard to match for statistical and philosophical sophistication in economics. A section of the reading list entitled "Instructor Bias" quotes Meehl: "Sir Ronald [Fisher] has befuddled us, mesmerized us, and led us down the primrose path. I believe that the almost universal reliance on merely refuting the null hypothesis as the standard method for corroborating substantive theories in the soft areas is a terrible mistake, is basically unsound, poor scientific strategy, and one of the worst things that ever happened in the history of psychology." Says Fraley to his graduate students, "I echo Meehl's sentiment."

Nonetheless the Task Force decided in short order that they would *not* be recommending the banning of significance testing. Reasons varied. For some it was to maintain freedom of inquiry and of method in science. A fine idea, but where, one might ask, were those voices for freedom and for rational method when the 1 percent/5 percent rule was codified in the *APA Manual* or when Melton was imposing his reign of 1 percent terror on *Experimental Psychology?* The task force did at least urge reporting effect size and in a meaningful context: "Reporting and interpreting effect sizes in the context of previously reported effects is essential to good research."[10] One step forward.

No Change

Fiona Fidler is a researcher in the Department of History and Philosophy of Science at the University of Melbourne. She and her coauthors have registered the significance error in psychology and medicine in the way we have for economics. They were not surprised that the recommendations of the task force have essentially led to "no change" in practice, remarking bitterly that "for a discipline that claims to be empirical, psychology has been strangely uninterested in evidence relating to statistical reform."[11] In a 2004 article entitled, "Editors Can Lead Researchers to Confidence Intervals, but They Can't Make Them Think," Fidler and another team of coauthors show that in psychology from the mid 1990s to the present only 38 percent "discussed clinical significance, distinguished from statistical significance" (Fidler et al. 2004b, 120). It was better, in other words, than the 1990s practice of the economists. But in light of the large investment made by the APA in changing the rhetoric of significance in psychology, the payoff was slight. Sixty-two percent said size *doesn't* matter.

In a major survey in 2000 of psychology and education journals by Vacha-Haase, Nilsson, Reetz, Lance, and Thompson the picture is worse. "Effect sizes have been found to be reported in between roughly 10 percent . . . and 50 percent of articles . . . notwithstanding either historical admonitions or the 1994 *Manual*'s 'encouragement' [to report effect sizes]."[12] The main exception would have been *Educational and Psychological Measurement,* edited by Bruce Thompson from 1995 to 2003. Thompson tried to attract articles to his journal that were devoted from start to finish to substantive significance. But he too saw little permanent progress. Recent issues of the journal, after Thompson, resemble on average the *American Economic Review* at its worst.

Typical is the experience of Philip Kendall in editing the *Journal of Consulting and Clinical Psychology*. In 1997 Kendall began to encourage authors to report on "clinical significance" and not merely the statistical significance of their results. In 1996 only about a third of the articles in the journal (fifty-nine in total) made *some* mention of clinical significance. The other two-thirds relied exclusively on statistical significance, similar to the *American Economic Review*. By 2000 and 2001 the situation had not much improved. Only 40 percent of the articles—4 percentage points more—drew a distinction between clinical and statistical significance (Fidler et al. 2004a, 619).

The Harvard Educational Review published in 1978 a devastating article by Ronald P. Carver against such whimsical disrespect for the substance of science. In 1993 Carver revisited the subject. "During the past 15 years," Carver wrote, "I have not seen any compelling arguments in defense of statistical significance testing." Meehl (1990) recently restated and reaffirmed the case against statistical significance testing. He too strongly condemned the whole tradition of using a rejected null hypothesis "as support for a theory" (1993, 287). The sizeless scientists, Carver too has found, do not have any compelling arguments. And yet they carry on and on.

Author Sloth

Around the same time an experiment in reporting strategy in the journal *Memory and Cognition* brought sad results. Despite the "requirement" by its editor, Geoffrey Loftus, that authors use error bars showing the confidence around their point estimates, less than half actually did so. Loftus was willing to impose the burden of the "new" reporting on himself. It has been said that he computed confidence intervals for more than a hundred of the articles submitted to the journal during his editorship. Although authors were asked officially in the back matter of the journal to do the work, and sometimes again in correspondence or in phone communication with the editor, hundreds reverted in their articles to the null-testing ritual and the sizeless stare. Maybe they do not know what a "confidence interval" is.

Psychometrics Lacks Power

Professor Savin: "What do you want, students?"
Iowa Graduate Students: "Power!"
Professor Savin: "What do you lack, students?"
Iowa Graduate Students: "Power!"

EUGENE SAVIN'S ECONOMETRICS COURSE,
UNIVERSITY OF IOWA, 1993

The cost of the psychological addiction to statistical significance can be measured by the "power function." Power asks, "What in the proffered experiment is the probability of correctly rejecting the null hypothesis, concluding that the null hypothesis is indeed false when it *is* false?" If the null hypothesis is false perhaps the other hypothesis—some other effect size—is true. A power *function* graphs the probability of rejecting the null hypothesis as a function of various assumed-to-be true effect sizes. Obviously the farther the actually true effect size is away from the null the easier it is going to be in an irritatingly random world to reject the null and the higher is going to be the power of the test.

Suppose a pill does in fact work to the patient's benefit. And suppose this efficacy is what the experiment reveals, though with sampling uncertainty. What you want to know—and are able in almost any testing situation to discover—is with how much power you can reject the null of "*no* efficacy" when the pill (or whatever it is you are studying) is in truth efficacious to such and such a degree. In general, the more power you have the better. You do not want by the vagaries of sampling to be led to reject what is actually a good pill.

There are reasons to quibble about this notion of power, as descended intuitively from Gosset and formally from Neyman and Pearson.

Sophisticates in the foundations of probability such as Savage and now Edward Leamer at UCLA have complained about its alleged objective certitude. Said Leamer to a 2004 assembly of economists, "[H]ypotheses and models are neither true nor false. They are sometimes useful and sometimes misleading" (Leamer 2004, 556). But this and other sophisticated complaints aside, power is considered by most statisticians—including Gosset and maybe Leamer—to provide a useful protection against unexamined null-hypothesis testing.

Power is, so to speak, "powerful" because hypotheses are plural and the plurality of hypotheses entail overlapping probability distributions. In a random sample the sleeping pill Napper may on average induce three extra hours of sleep, plus or minus three. But in another sample the same scientist may find that the same sleeping pill, Napper, induces two extra hours of sleep, plus or minus four (after all, some sleeping pills contain stimulants, causing negative sleep). The traveler would like to know from her doctor before she takes the pill exactly how much confidence she should have in it. "With what probability can I expect to get the additional two or three hours of rest?" she reasonably wants to know. "And with what probability might I actually get *less* rest?"

Without a calculation of power, to be provided by the psychometricians, she can't say. Calculators of Type I error pretend otherwise: following the practice of R. A. Fisher, they act as if the null hypothesis of "no, zero, nada additional rest" is the only hypothesis that is worthy of probabilistic assessment. They ignore the other hypotheses. They tell the business traveler and other patients: "Pill Napper is statistically significantly different from zero at the 5 percent level." To which their better judgment—their Gosset judgment—should say, "So What?"

Power is, mathematically speaking, a number between zero and 1.0. It is the difference between 1.0 (an extremely high amount of power, a good thing) and the probability of an error of the second kind (a bad thing). The error of the second kind is the error of accepting the null hypothesis of (say) zero effect when the null is in fact false, that is, when (say) such and such a positive effect is true. Typically the power of psychological research is called "high" if it attains a level of .85 or better. (This, too, is arbitrary, of course. A serious study with a loss function may not accept a hard and fast rule.) High power is one element of a good rejection. If the power of a test is low, say, .33, then the scientist will two times in three accept the null and mistakenly conclude that another hypothesis is false. If on the other hand the power of a test is high, say, .85 or higher, then the scientist can be reasonably confident that at minimum the null hypothesis (of, again, zero effect if that is the null chosen) is false and that therefore his rejection of it is highly probably correct.

If the "null" is "no efficacy at all, when I would rather find a positive effect of my beautiful sleeping-pill theory," too often rejecting the null without consideration of the plurality of alternatives is the same thing as doing bad science and giving bad advice. It is the history of Fisher significance testing. One erects little "significance" hurdles, six inches tall, and makes a great show of leaping over them, concluding from a test of statistical significance that the data are "consistent with" ones own very charming hypothesis.

A good and sensible rejection of the null is, among other things, a rejection *with high power*. If a test does a good job of uncovering efficacy, then the test has high power and the hurdles are high not low. The skeptic—the student of R. A. Fisher—then is properly silenced. The proper skeptic is a useful fellow to have around, as Descartes observed. But in the Fisherian way of testing a null as if absolutely, by the 5 percent criterion, the skepticism is often enough turned on its head. It is in fact gullibility posturing as skepticism. That is, in denying the plurality of overlapping hypotheses, the Fisherian tester asks very little of the data. She sees the world as Annie Dillard once did, through the lens of one hypothesis—the null.

To put it another way, power puts a check on the naïveté of the gullible. He, too, a faithful fellow, can be useful, as Cardinal Newman observed. But the failure to detect a significant difference between two sleeping pills, say, Napper and its market competitor, Mors, does not mean that a difference is not there in God's eyes. A Fisher test of significance asks what the probability is of claiming a result when it is *not* really there, that is, when the null hypothesis is true: no efficacy. Power protects against undue gullibility, then, an excess of faith. Power is a legitimate worry.

Gosset discovered the legitimate worry, we have said, in his letter of May 1926, pointing out to Egon Pearson that the significance level trades off against power, still to be named. The confidence we place in Student's *t* depends, Student said, other things equal, on the probability of one or more relevant "alternative hypotheses" perhaps more true. Naively accepting the singular null hypothesis involves a loss—"but *how much do we lose?*".[1] In 1928, and then more formally in 1933, Neyman and Pearson famously operationalized Gosset's improvement of his own test.

Yet power is usually ignored in psychometric practice. It is wrong to be too gullible, granted. But it is also wrong to be too skeptical. If you protect yourself from gullibility in thinking an antidepressant is efficacious, you will avoid the embarrassment and cost of recommending peach pits when they don't work. But if you don't *also* protect yourself from excessive skepticism, by getting sufficient power, you will *not* avoid the other cost—of dead patients who might have been saved by a pill that does work. You will have set the hurdles too high rather than too low.

After Fisher, few scientists get it. Ziliak had the unusual experience in 2006 of being dumped by a family physician at a good hospital in Chicago. The ostensible reason? She grew angry at his question about the power of a pill she had prescribed. She couldn't understand.

Setting the height of the statistical hurdles involves a scarcity, just as the setting of real hurdles does. Holding sample size constant, seeking low (mistaken) skepticism—high statistical significance—has the inevitable opportunity cost of higher (mistaken) gullibility. For a given sample size, power is a declining function of significance level (fig. 12.1). This makes sense: the more area under the bell curve you want to yield to your null experiment (making rejection of the null more difficult by lowering the level of Type I α-error), the more you encroach into the probability distributions—the bell curves—of adjoining hypotheses.

But high power is no permashield against other kinds of oomph-ignoring errors rife in the statistical sciences. To estimate the power function one needs to define among other things a domain of relevant effect sizes different from the null. And that decision is about oomph. The 2003 article on Vioxx is proof of what can go wrong when the oomph of the test is not attended to, even though the power of the test is. "A sample size of 2,780 patients per treatment group," the authors of the infamous study said, "was expected to provide 90 percent power to detect a difference of 2 percentage points between treatments for the primary safety variable" (Lisse et al. 2003, 541). But as we have seen the authors did not estimate the power of their test to reject the hypothesis of no harmful cardiac effect between Vioxx and naproxen. Pretending to be excessively gullible, they ignored a 8-to-1 cardiac damage or death ratio, a magnitude or "safety variable" of some importance.

How to Get Powerful

Mosteller and Bush (1954) seem to be the first to have assessed the amount of statistical power in the social sciences. The psychologist Jacob Cohen was the first to conduct a large-scale, systematic survey of it in psy-

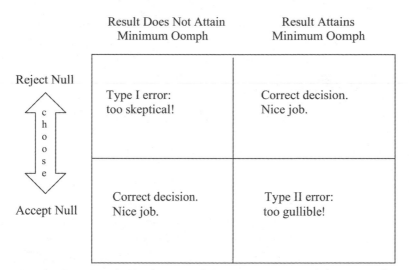

Fig. 12.1. Does your decision have oomph? Knowing your power helps. Power asks, holding my degree of skepticism constant, how gullible am I? One error trades off the other. Most Fisher tests reject with a power of less than 50 percent—as good as your local psychic.

chology proper (1962). Cohen surveyed all seventy articles published in the *Journal of Abnormal and Social Psychology* for the year 1960, excluding minor case reports, factor-analytic studies, and other contributions for which the calculation was impossible.

To calculate a power function one needs a random sample, a fixed level of significance (Type I error of, say, .05), and one or more measures of effect size different from the null and from the result obtained. The effect size is the assumed efficacy in God's eyes, so to speak, which you should be uncovering. (In the face of such language one can sympathize with the secular pragmatism of a Leamer or a Savage.) If you have a very large sample, there is a pseudo-problem of power. With $N = 10,000$ even weak effects will show through a cloud of skepticism. Everything will be significant, and with high power, though in that case the significance, or the power, of an effect is not itself much of an accomplishment. If you are a Fisherian, the fact of a large sample becomes your problem. You're deluded, thinking you've proved oomph before you've considered what it is.

In psychology, Cohen noted in his survey, as often and more alarmingly than in medicine, few in 1960 reported the effect size they had found. A reader could not therefore, even if she had wanted to, estimate the power of their tests against the alternative effect sizes. Power estimation requires effect sizes. So in his large-scale survey of power Cohen had to *stipulate* the effect sizes, assigning what seemed to him small, medium, and large magnitudes for each case. It was quite a task. Having done a little of this sort of thing ourselves (and on desktop personal computers in 2005, not on old Frieden mechanical calculators in 1962) we stand amazed at his scientific energy. For articles using *t*-tests Cohen assigned .25, .50, and 1.00 standard deviation units to stand for small, medium, and large effect. For articles using Pearson correlation coefficients he used .20, .40, and .50 standard deviation units to stand for small, medium, and large effect.

Cohen's standard, alas, is a merely statistical one. On such heterogeneous subjects as one finds in the *Journal of Abnormal and Social Psychology*—from the relation of a medical treatment to paranoid schizophrenia to the relation of mother and son to sexual fetishism—a different investigator might well have divided up the regions of effect size in a different way. Cohen would have had to be expert in every subliterature in psychology to judge for each the relevant standard of effect. For some phenomena a 0.20 standard deviation change may produce a *large* clinically or ethically important effect on the dependent variable, such as anxiety or crime rate reduction. Cohen himself, fully aware of the issue, suggested in 1969 a downward revision of his 1962 effect sizes (Cohen 1969).

Still, Cohen's standard of effect size is a good deal better than nothing, and has the advantage of being easily replicable. And for our current point it suffices: Cohen established a measure of the largeness of effect that allows actual calculations of power. The authors of the original articles did not, you understand.

Cohen's three assumptions about effect size gave him three levels of power for each of the 2,088 tests of significance in the seventy articles—notice that even in 1960, long before electronic computers, the average article in the *Journal of Abnormal and Social Psychology* was exhibiting about thirty significance tests. Thirty tests per article. The price per test fell dramatically in the next few decades, and, as an economist would expect, the number of tests per article correspondingly ballooned into the hundreds (not all of them published).

From the large-scale survey Cohen reckoned that the power in detecting "large" effects was about .83. So the probability of mistakenly rejecting a treatment having a "large" effect is of course 1.00 minus the .83 power or 17 percent. That seems satisfactory, at any rate for a moderate loss from excessive skepticism. On the other hand, if you were dying of cancer you might not view a 17 percent chance of needlessly dying as "satisfactory," not at all. You might well opt for peach pits. It always depends on the loss, measured in side effects, treatment cost, death rates. The loss to a cool, scientific, impartial spectator will not be the same as the loss to the patient in question. In 1933 Neyman and Pearson said, "[H]ow the balance should be struck" between Type I and Type II errors "must be left to the investigator."[2] That formulation is progress over the sizeless stare. But it would seem that a better formulation in medicine is that it "must be left to the patient, friends, and family."

At smaller effect sizes, Cohen found, power faded fast. For the effects assumed to be in God's eyes "medium" or "small" the power Cohen found was derisory. It was, averaging one article with another, about .48 for medium-size effects and only .18 for small. That is, for a small, 0.25 standard deviation unit departure from the null hypothesis, the wrong decision was made 1.00 minus 0.18 or 92 *percent of the time*. (Cohen in 1969 redid the power calculations at lower effect sizes and got about the same results.)

The pattern was the same in similarly large-scale studies conducted by Sterling (1959), Kruskal (1968b), and Gigerenzer et al. (1989). In fact dozens of additional surveys of power in psychology have been performed on the model of Cohen's original article. Rossi summarizes the findings: "The average statistical power for all 25 power surveys [including Cohen's] was .26 for small effects, .64 for medium effects, and .85 for large effects and was based on 40,000 statistical tests published in over 1,500 journals" (1990, 647). For example, years later, Sedlmeir and Gigerenzer (1989) also surveyed the power of research reported in the *Journal of Abnormal Psychology*. Using Cohen's original definitions of effect size, they found mean power values of .21, .50, and .84—in other words, nearly the same as Cohen found for small, medium, and large effect sizes decades earlier.

The Power of Rossi

A power study by Joseph Rossi (1990) should have been crushingly persuasive. Rossi calculated power for an astonishing 6,155 statistical tests in 221 articles. The articles had been published in the year 1982 in three psychology journals, the *Journal of Abnormal Psychology, Journal of Consulting and Clinical Psychology,* and *Journal of Personality and Social Psychology.* We again stand in awe: would that critics of the idiocy of null-hypothesis significance testing in economics—not excepting Ziliak and McCloskey—had such scientific energy. Using Cohen's effect sizes,

Rossi found power to detect large, medium, and small effects of .83, .57, and .17. He calculated power for 1,289 tests in the *Journal of Abnormal Psychology,* 2,231 in the *Journal of Consulting and Clinical Psychology,* and 2,635 in the *Journal of Personality and Social Psychology.* The conclusion: "20 years after Cohen conducted the first power survey, the power of psychological research is still low" (646). And almost twenty years after Rossi it is still equally low.

Usually, as we have seen, the statistical test is not of an efficacy of treatment so much as *inefficacy,* that is, a null of No Effect from which the psychologist wants to conclude that there *is* an effect. Either way, low power is a scientific mistake. As Rossi writes, "[I]f power was low, then it is reasonable to suggest that, a priori, there was not a fair chance of rejecting the null hypothesis and that the failure to reject the null should not weigh so heavily against the alternative hypothesis" (1990). That's to put it mildly: the six-inch hurdles are lined up and the six-foot scientist courageously leaps over them. The scandal of low power in the social sciences should bring their practitioners to some humility. Yet Fisherian testers are *very* proud of their rejections of the null and very willing to impose conformity in leaping aristocratically over them. By contrast, Rossi recommends Gosset-like expressions of "probable upper bounds on effect sizes." We would add only that "probable *lower* bounds on effect sizes" are also needed (cf. Würtz 2003 and Leamer 1982).

A "real" index of Type I error might help: "real" Type I error is the ratio of the *p*-value to the power of the test.[3]

real Type I error = empirical *p*-value/empirical power of the test

An alleged $p = .05$ will turn out actually to be an alarming "real" *p* of .20 if the power of the test is only .25. An alleged $p = .10$ is really .33 if the power of the test is .30. Recall in the power studies how often this is the case for small effect sizes. Real *p*-values in excess of .30 are the norm. Reporting the real level of Type I error has the advantage of allowing the reader to approximate how many "false" rejections of the null will occur for every "true" or correct rejection.

According to Rossi, the real rate of false rejections in psychology is grim: "More than 90% [of over six thousand] of the surveyed studies had less than one chance in three of detecting a small effect"—very far above Fisher's 5 percent error claimed (1990). Psychologists need to know that the real rate of false rejection is for small effect sizes at best .05/.17, or about 29 percent, and for medium-sized effects .05/.57, or about 9 percent. The same is true of economics and its imitative younger brother, political science. In other words, a 5 percent significance test is actually a vaguer 9 or 29 percent significance test. So much for precision.

The problem of making a dull tool appear sharp is rife. Robert Shiller, a leading financial economist, wrote in his "The Volatility of Stock Market Prices" in *Science* that "the widespread impression that there is strong evidence for market efficiency may be due just to a lack of appreciation of the low power of many statistical tests" (1987). Can other scientists claim to know as much about the statistical power of their field as the psychologists do? Very few. In economics we think of Zellner, Horowitz, Eugene Savin, and Allan Würtz. In heart and cancer medical science, Jennie Freiman et al. (1978) know their power. A few biologists and ecologists can claim to know their power, too, for example, Anderson et al. (2000). Most statistical scientists do not.

Designing experiments to find the maximal and minimal effect size is a better way to get powerful results and to keep the focus where it should be: on effect size itself. As Gosset argued at the Guinness factory, so Rossi says:

> Increasing the magnitude of effects may be the only practical alternative to expensive increases in sample size as a means for increasing the statistical power of psychological research. We tend to think of effect size (when we think of it at all) as a fixed and immutable quantity that we attempt to detect. It may be more useful to think of effect size as a manipulable parameter that can, in a sense, be made larger through greater measurement accuracy. This can be done through the use of more effective measurement models, more sensitive research designs, and more powerful statistical techniques. Examples might include more reliable psychometric tests; better control of extraneous sources of variance through the use of blocking, covariates, factorial designs, and repeated measurement designs; and, in general, through the use of any procedures that effectively reduce the "noise" in the system. (1990, 654)

Those sound like good ideas for a science. Better than mushy *p*s.

13

The Psychology of
Psychological Significance Testing

A significance test is more likely to suggest a difference than is Jeffreys' [1939] method. This may partly account for the popularity of tests with scientists, since they often want to demonstrate differences. It would be interesting to know how many significant results correspond to real differences. It is also interesting that many experimentalists, when asked what 5 percent significance means, often say that the probability of the null hypothesis is 0.05. This is not true, save exceptionally. In saying this they are thinking like Jeffreys but not acting like him.

D. V. LINDLEY 1991, 12

Why have psychologists been unwilling to listen? One reason seems to be insecurity in a so-called soft or subjective field. Recall even the learned Paul Meehl, a psychological scientist as well as a philosopher, speaking of his own field as "soft." The "hard/soft" dichotomy is surely a poor one for any science. It does not acknowledge the hardness of Greek contrary-to-fact conditionals in a soft field like classics or the softness of linked index numbers in a "hard" field like economics. Like soft and hard, *subjective* and *objective* are laymen's not philosopher's terms and no better for it. The ontology and epistemology implied are dubious. Deciding what is hard and what soft, objective and subjective, in a chain-weighted adjustment of the gross domestic product (GDP) price deflator is an irrelevant diversion from the central scientific question: what in the current state of the science persuades?

Fisher-significance is a manly sounding answer, though false. And one can see in the dichotomy of hard and soft a gendered worry, too. The worry may induce some men to cling to Significance Only. Barbara Laslett, for example, has written persuasively of the masculine appeal of quantification in American sociology after 1920 (2005). By the early 1920s the percentage of articles published in sociology journals that employed statistical methods exceeded 30 percent (Ross 1991, 429). "Statistical methods are essential to social studies," Fisher wrote in the first edition of *Statistical Methods for Research Workers* (1925a). "It is principally by the aid of such methods that these studies may be raised to the rank of sciences" (2). Hardboiled-dom was the rule such fields used to raise themselves to a 5 percent science.

Around 1950, at the peak of gender anxiety among middle-class men in the United States, nothing could be worse than to call a man soft. Consider it. "For this"—the era of war, of depression, observed Saul Bellow—"is an era of hardboiled-dom," and was, too, the era of Fisher and Yates, Hotelling and McNemar. The "code . . . of the tough boy—an American inheritance, I believe," said Bellow, "from the English gentleman—that curious mixture of striving, asceticism, and rigor, the origins of which some trace back to Alexander the Great—is [in the 1940s] stronger than ever. . . . They [e.g., the Fisherian hard-boiled] are unpracticed in introspection, and therefore badly equipped to deal with opponents whom they cannot shoot like big game or outdo in daring" (1944, 9).

THE HARDNESS OF THE SOFT SCIENCES

Psychology is anyway nothing like as soft as is sometimes believed. Psychologists have long employed macho statistics and in the beginning at a level akin to the commonly used techniques in English biometrics. In Germany the experimental psychologists began to use statistics as early as the middle of the nineteenth century.[1] Wilhelm Wundt and especially Gustav Fechner, the father of "psychophysics," were the first to walk philosophy of mind down the wooden stairs of German metaphysics and into the counting room of empirical perception. Wundt's and Fechner's laboratories worked on "applications" only. Theory would remain, to the new experimentalists, deterministic. Yet their demarcation of theory and practice—of ideas and applications—was friendly toward statistical testing and classical inference. "A frequency interpretation grounded this work," writes the historian of statistics David Howie, "since the categories of perception were defined on a scale marked by the ratio of an individual's repeated assessments of the physical stimulus under consideration."[2] To the experimentalists a statistic could measure the degree of the grounding. Their tools were in the 1870s understandably primitive but in no sense soft. Fechner himself employed means, standard deviations, and the occasional coefficient of variation.

By the early twentieth century Americans and Europeans alike began to learn, especially from the London statisticians, the ways and means of hard-boiled "testing." American psychologists, like many of America's human and life scientists, were especially open to it. Measurement felt hard. The biometric methods of Karl Pearson were introduced to psychologists by, it seems, the Columbia professors F. S. Chapin and Franklin Giddings. *Psychometrica* was founded by L. L. Thurstone, a University of Chicago "psychophysicist," as he called himself, who began in the late 1920s to use Fisher's inferential techniques.[3] Fisher had given Thurstone and others the scientific legitimacy they sought. Significance testing and quantitative methods generally were promoted early and late by the Social Science Research Council. Chapin, Giddings, Thurstone, and others wanted to free themselves from what they took to be that mere "opinion and crankery" of nonquantified fields.[4]

A wider cultural "trust in numbers" had triumphed, and the life and human sciences, including psychology, would trumpet their new trust.[5] In 1910 the Flexner Report on medical education advocated a scientific medicine and a monopoly of a small group of medical schools. In 1915 Flexner told a gathering of social workers that if they wanted a formula for professional success—standardization of procedures, consensus in decision making, monopolization of goods and services, higher salaries—they should follow the model he had designed for medicine.[6] They did. Flexnerian professionalization—inspired by the positivism of Karl Pearson—is one reason that early-twentieth-century social workers, drawn initially to problems of poverty and human rights, ended up in the service of bourgeois values and spying for the state. A rather similar pattern is found in nursing, with a lag.

In this context the 5 percent science was promoted by the new leaders of quantitative psychology and education. European humanists can score themselves by how many generations they are removed from Hegel—that is, in being taught by a teacher who was taught by a teacher who was taught by a teacher who was taught by Hegel at the University of Berlin. Likewise, statisticians can score themselves by how many generations they are from Fisher. Quinn McNemar, for example, of Stanford University, was an important teacher of psychologists who had himself studied statistical methods at Stanford with Harold Hotelling, the chief American disciple of Fisher. Hotelling had worked directly with Fisher. McNemar then taught L. G. Humphreys, Allen Edwards, David Grant, and scores of others. As early as 1935 all graduate students in psychology at Stanford, following the model of Iowa State, were required to master Fisher's crowning achievement, analysis of variance. Already by 1950, Gigerenzer et al. reckon, about half of the leading departments of psychology required training in Fisherian methods (1989, 207).

Even rebels against Fisher were close to him, starting with Gosset himself. Palmer Johnson of the University of Minnesota studied with Fisher in England, though he later had the bad taste to write articles with Fisher's erstwhile colleague and eternal enemy Jerzy Neyman, whom Fisher had cast into outer darkness. George Snedecor, an agricultural scientist at Iowa State University at Ames, was a cofounder of the first department of statistics in the United States. His important book *Statistical Methods* was influenced directly by Fisher himself, who somewhat surprisingly was in the 1930s a visiting professor of statistics at Iowa State. One can think of the Iowa schools then as one thinks of London's Gower Street in the 1920s and 1930s—a crucial crossroads of statistical methods and training. In a eulogy for S. S. Wilks, a student in the late 1920s of Henry L. Rietz and Allen T. Craig at the University of Iowa, Frederick Mosteller said that Iowa was then "the center of statistical study in the United States of America" (1964, 11). (Craig co-authored a famous text with Robert Hogg, whom Ziliak took lectures from.) Rietz, Craig, and Wilks worked closely with Fisher. E. F. Lindquist, the American leader of standardized testing for educators, also of the University of Iowa, was deeply influenced by Snedecor. Lindquist invented the Iowa Test of Basic Skills for schoolchildren. He too spent time with the great man.

Some psychologists knew about the work of Neyman and Pearson and some even about that of the Bayesian Harold Jeffreys. But textbook authors, editors, and teachers—inspired by Fisher's promise of raising their fields to the level of hard science—helped Fisher win the day. Statistical education narrowed at the same time as it spread. Decision theory and inverse probability, and Gosset's views on substantive significance, alternative hypotheses, and power, were pushed aside. Too introspective for the hard-boiled.

J. P. Guilford's influential *Fundamental Statistics in Psychology and Education* (1942) decided that power was, in his words, "too complicated to discuss."[7] In 2004 an influential textbook writer, a psychologist, told

Gigerenzer that he regretted leaving out power. He noted that in the first edition of his successful textbook he had discussed Bayesian and decision-theoretic methods, including power. "Deep down," he confessed, "I am a Bayesian." But in the second and subsequent editions the notions of power and decision theory and the costs of decisions, both Bayesian and Ney-man-Pearson, vanished. The "culprits," Gigerenzer believes, were "his fellow researchers, the university administration, and his publisher."[8]

Fisher Denies the "Cost" of Observations

During the 1940s and 1950s and 1960s among a tiny group of sophisticates the Bayes and Neyman-Pearson approaches were the gold standard. But at lower levels of statistical education Bayes and Neyman-Pearson, as we have said, were seldom presented. Gosset's economic approach to statistics, picked up later by Savage and Blackwell and Zellner, was at mid-century invisible. Fisher realized that acknowledging power and loss functions would kill the unadorned significance testing he advocated and fought to the end, and successfully, against them (Fisher 1955, 1956). "It is important that the scientific worker *introduces no cost functions for faulty decisions,*" Fisher wrote in 1956, "as it is reasonable and often necessary to do with an Acceptance Procedure" in something so vulgar as manufacturing.

> To do so would be to imply that the purposes to which new knowledge was to be put were known and capable of evaluation. If, however, scientific findings are communicated for the enlightenment of other free minds, they may be put sooner or later to the service of a number of purposes, of which we can know nothing. The contribution to the improvement of Natural Knowledge, which research may accomplish, is disseminated in the hope and faith that, as more becomes known, or more surely known, a great variety of purposes by a great variety of men, and groups of men, will be facilitated. No one, happily, is in a position to censor these in advance. As workers in Science we aim, in fact, at methods of inference which shall be equally convincing to all freely reasoning minds, entirely independently of any intentions that might be furthered by utilizing the knowledge inferred. (102–3)

Fearful of the growing attraction of decision theory among the mathematical sophisticates, Fisher tried to identify Deming-type and Neyman-Pearson-type decisions with, as Savage put it mockingly, "the slaves of Wall Street and the Kremlin."[9] With Deming and his teacher Shewhart in mind, Fisher wrote in 1955, "In the U.S. also the great importance of organized technology has I think made it easy to confuse the process appropriate for drawing correct conclusions [by 'correct' he means of course his own sizeless stare at Type I error and his logically flawed *modus tollens*], with those aimed rather at, let us say, speeding production, or saving money" (70). Thus "Wall Street." Notice the sneer by the new aristocracy of merit, as the clerisy fancied itself. Bourgeois production and money making, Fisher avers, are *not* the appropriate currencies of science.

Nor the Kremlin. Neyman was a Polish Catholic but *raised in Russia*. A progressive, *he was active in the American civil rights movement,* and tried unsuccessfully for years to get his Berkeley colleagues to hire his friend, the mathematician David Blackwell (b. 1919), whose chief defect as a mathematician was the color of his skin.[10] Ah-hah. (In 1949 Blackwell published an anti-Fisher article, with Kenneth Arrow and Martin Girshick, on optimal stopping rules in sequential sampling, a "loss function" idea inspired by Wald and Bayes.)[11] Blackwell credited Savage, with whom Blackwell had worked at the Institute for Advanced Study, for showing him the power of the Bayesian approach. "Jimmy convinced me that the Bayes approach is absolutely the right way to do statistical inference," he said.[12]

Fisher viewed science as something quite distinct from "organized technology" and the questionable social purposes of the left wing.

> I am casting no contempt on acceptance procedures [he continued], and I am thankful, whenever I travel by air, that the high level of precision and reliability required can really be achieved by such means. But the logical differences between such an operation and the work of scientific discovery by physical or biological experimentation seem to me so wide that the analogy between them is not helpful, and the identification of the two sorts of operation is decidedly misleading. (1955, 69–70)

To which Savage would later reply: "[i]n the view of a personalistic Bayesian like me, the contrast between behavior and inference is less vivid than in other views. For in this view, all uncertainties are measured by means of probabilities, and these probabilities, together with utilities [or natural selection or whatever your currency is measured by], guide economic behavior, but the probability of an event for a person (in this theory) does not depend on the economic [or psychological] opportunities of the person" (1971a, 465). And yet, notes Savage, "almost in the same breath with criticism of . . . decision functions, Fisher warns [in 1958,

274] that if his [5 percent] methods are ignored and their methods used a lot of guided missiles and other valuable things will come to grief" (Savage 1971a, 465).

In the 1950s the sharpest of Fisher's stabs were directed at Neyman and induced an immediate response (Neyman 1956): "As Sir Ronald remarks correctly, merely from the specification of the null hypothesis, the probabilities of errors of the second kind are certainly not calculable. However, the main point of the modern theory of testing hypotheses [of Gosset, Egon, Neyman, Wald, Savage, Lindley, and Blackwell] is that, for a problem to make sense, its datum must include not only a [null] hypothesis to be tested, but . . . the specification of a set Ω of alternative hypotheses that are also considered admissible. . . . With this kind of datum," Neyman concludes, "the probabilities of errors of the second kind are certainly calculable and, in fact, a considerable number of tables of such probabilities, or of their equivalents, namely of power functions, are now available" (290).

Yet textbook writers have launched a different guided missile, concocting since the 1950s a hodgepodge of Fisher and his enemies. Early on in an elementary statistics or psychometrics or econometrics book there might appear a loss function—"what if it rains the day of the company picnic?" But the loss function disappears when the book gets down to producing a formula for science. In the more advanced texts the discussion of power and decision theory comes late or not at all. At both levels, elementary and advanced, the modern theory of testing hypotheses is marginalized. The names in the competition, Gosset to Lindley, have long since dropped out. The hodgepodge, as Gigerenzer et al. (1989) have noted, is introduced anonymously, as if there is only one way of testing and estimation, and Fisher's is it (cf. Efron 1996). There is no god of profitable scientific action, but Sir Ronald is his prophet.

A philosopher of science, Deborah Mayo (1999), has recently entered the debate in favor of a hybrid of Neyman-Pearson decisions and Fisher's rule of significance for accepting an experimental result. Her *Error and the Growth of Experimental Knowledge* attempts to steer scientists toward a more systematic analysis of "errors," a Popperian direction we salute. Like Gosset (of whom Mayo is apparently unaware) and the Bayesian Harold Jeffreys, she "came to see" how "statistical methods . . . enable us, quite literally, to learn from error."[13]

But Mayo places too much faith in the ability of tests of significance to guide error-based, experimental learning. Our point is that such a guide is useless, and therefore will not direct science to the *right* correction of its errors. Central to her "radically different" (1999, 13) notion is "The falsification of statistical claims . . . by standard [for her, the standard Neyman-Pearson-Fisher hybrid] statistical tests" (13). A strength of her project is, she says, "fundamental use of this [Neyman-Pearson] approach, albeit reinterpreted, as well as of cognate methods [e.g., Fisherian tests]. My use of these methods," she says, "reflects their actual uses in science" (16).[14] Alas. If one turns to Mayo's discussion on what constitutes a "severe test" of an experiment, one finds only sizeless propositions, with loss or error expressed in no currency beyond a scale-free probability. Her notion of a severe test would seem beside the point to her muse, Egon Pearson, and especially to Egon's muse, William Sealy Gosset (178–87, 460).

A better approach to error-based learning that keeps both statistical hypothesis testing and experimental control at its center, as Mayo desires, would put Gosset's notion of *net pecuniary value* in its center. A notion of a severe test without a notion of a loss function is a diversion from the main job of science, and the cause, we have shown, of error. As Gosset said to Karl Pearson in 1905, "[I]t would appear that in such work as ours the degree of certainty to be aimed at must depend on the *pecuniary advantage to be gained by following the result of the experiment, compared with the increased cost of the new method, if any, and the cost of each experiment.*" If human life, not money, is what is at stake in the experiment, then loss of life should define the "error" function (cf. Freiman et al.) And so forth, across the sciences. In the end what is at stake is the persuasion of other scientists, even in cases like astronomy in which no cost or life hinges on a decision.

Considering the size of the intellectual investment, many psychologists worry that psychology has not learned much from its statistical revolution. We believe the same can be said of econometrics. In both places the problem can be put as the low statistical power of the tests, coming from the one-sided devotion to Fisherian procedures. As Eugene Savin asks of his students in econometrics at the University of Iowa, "What do you *want,* students?" "Power!" he trains them to enthusiastically shout. "And what do you *lack,* students?" "Power!" Savin's way of putting it may seem to an outsider a claim that power functions offer a full solution. It does not, and Savin does not think it does.

The big ideas of psychological theory, one can argue, have not been shaped by Fisherianism. As in economics and biology and physics, the big ideas come chiefly from nonstatistical sources. "Wolfgang Kohler derived his Gestalt laws," Gigerenzer et al. observe of the classic period of psychological theory, "Jean Piaget his theory of intellectual development, B. F. Skinner his principles of operant conditioning, and Sir Frederic Bartlett his theory of memory, all without a statistical method of inference."[15] William James, John Dewey, and, to give a recent example, Howard Gardner, did not depend on Student's *t*. And yet, after Fisher, IQ testers and the students of personality still base their claims to science on statistical measures of correlation and fit. Educational psychology was therefore particularly vulnerable. Its flagship journal, the *Journal of Educational Psychology,* has been saturated by significance testing of the Fisher type since the early 1940s.

I'm OK, You're a Bayesian

There's a contradiction here, well known to Mayo, Gigerenzer, and other philosophers of knowledge, and particularly evident in the literature concerning the psychology of learning: anyone modeling human learning adopts something like a Bayesian framework. Cognitive psychologists and philosophers of mind ascribe a "testing propensity" to the human mind itself, claiming that it is Bayesian or decision theoretic in nature. Piaget, for example, spoke of the child as if she were a "scientist." The child learns as scientists do, conditioning present actions on the most recent information. Max Frisch characterized humans as "gamers," *Homo ludens,* as the Dutch historian Jan Huizinga famously called us. It is an idea a Bayesian game theorist could readily agree with. By the 1960s the cognitive psychologists, operations researchers, and a few economists had begun to argue that "rational" economies were tacitly or explicitly Bayesian.[16] Others believed the mind to be a "Neyman-Pearson detector" or "observer." Human action, they said, in notable contrast with Fisherian docility, was necessary, and possible—independent of "belief." The mind's internal detector simply

> adopts the decision goal of maximizing the hit rate (i.e., the correct detection of a signal) for a given false alarm rate (i.e., the erroneous "detection" of a signal). This view of perceptual processing was in analogy to Neyman and Pearson's concept of "optimal tests," where the power (hit rate) of a test is maximized for a given type-I error (false alarm rate). (Gigerenzer et al. 1989, 214)

Human subjects, they said, know things already, prior to the experiment, of course. They learn from acting and by being, by learning and by doing, and update their beliefs about the world accordingly. Of course.

But—here's the contradiction—when such Bayesians or quasi-Baye-sians went to do the statistical analysis on their observations or experi-ments, they used *Fisher* tests of significance, ignoring their own Bayesian framework of beliefs and learning by doing. Some few have on the con-trary adjusted psychology to Fisher, such as Harold H. Kelley (1967), proposing that "the mind attributes a cause to an effect in the same way as behavioral scientists do, namely by performing an analysis of variance and testing null hypotheses" (214). Kelley's analysis-of-variance model of causal attribution generated for twenty years a large amount of re-search in experimental social psychology. One doubts that it makes sense. But most in psychology were theoretical Bayesians and Fisherian testers, a contradiction. Gigerenzer et al. conclude that "the reasons for this double standard seem to be mainly rhetorical and institutional rather than logical" (1989, 233).

But Fisherians Survive on the Illogic of Neopositivism

The schizoid rhetoric is that of neopositivism. One reason for the success of the Fisherian program against more logical alternatives, such as Bayesianism or Neyman-Pearson decision theory or Gosset-Savage econo-mism, is that the Fisherian program emerged just as neopositivism and then falsificationism emerged in the philosophy of science. It would have fallen flat in philosophically more subtle times, such as those of Mill's *Sys-tem of Logic Ratiocinative and Inductive* (1843) or Imre Lakatos's *Proofs and Refutations* (1976). No serious philosopher nowadays is a positivist, no serious philosopher of science a simple falsificationist. But the philo-sophical atmosphere of 1922–62 was perfect for the fastening of Fisher's grip on the sizeless sciences.

Fisher recommended that the investigator focus on falsifying the null. If you suppose that correlation of infant habituation to novel stimuli, measured by eye gaze, predicts childhood IQ "remarkably well (0.7, cor-rected for unreliability)," you "test" it by asking what is the probability of a result as high as 0.7 if the null were true, namely, if zero correlation were true (Gottfredson 1996, 21). If the probability is suitably small you have falsified the null hypothesis and conclude triumphantly that the maintained hypothesis, namely, 0.7, must therefore be true.

This is the method recommended by Popperian philosophy of science since the 1930s, after Popper's self-declared "killing" of the confirmation-ism of Vienna logical positivism. Both versions of positivism—Popper in truth modified it rather than killed it—hang on the hypothetico-deductive view of science. The notion is that your theory, T, can be stated as axioms, A, which imply hypotheses, H, which in turn imply observations, O. The theory is tested, says the hypothetico-deductive philosopher, by measuring O and then working back up the chain of implication through H and A to the grand T.

In other words, science is supposed to be summarizable as a *logical* system—hence "logical" positivism and the hypothetico-*deductive* view. Suppose H_0, the hypothesis of no connection whatever between an infant's ability to shift her gaze and the IQ of the little girl she grows into, implies some observations in the world, O. Symbolically, $H_0 \Rightarrow O$. Then, by what is known in logic as the *modus tollens,* it follows strictly that not-$O \Rightarrow$ not-H_0. If ignition in the presence of gasoline implies combustion, then a *lack* of combustion strictly implies a lack of ignition. If independence between gaze shift and later IQ implies the presence of a low correlation between the two, then a *lack* of a low correlation strictly implies a lack of independence.

So, if you observe a correlation between gaze shift and later IQ (of, say, 0.7), which is very unlikely to occur in a world in which the null is true, then you have observed the probabilistic version of not-O. Therefore by *modus tollens* you know that the null is (probabilistically) falsified, that is, probably not-H_0 is true. Consequently—*this step is not valid*—your alternative to the null is true. Consequently, H is true and so is A and so is T. Consequently, you can with assurance assign children to streams in school and university and life on the basis of their gaze shift as infants. Thus the mismeasurement of man.

You Can Abandon Falsification and Still Be Scientific

Falsificationism has retained a grip on scientists with a little philosophical learning ever since it was first articulated by Popper. The notion was not in English firmly tied to Popper's name until his PhD dissertation, published in 1935 as *Logik der Forschung,* was translated into English as *The Logic of Scientific Discovery* in 1959. Most scientists nowadays, if they have philosophical ideas, reckon that they are Popperians. No wonder, since Popper portrayed the scientist as a hero courageously facing up to his own refutation within a system of strict logic.

Falsificationism and the hypothetico-deductive view and logical positivism, and therefore Fisherianism, however, have had a flaw all along. It is that they are illogical, erroneous deductions. The flaw in logic was pointed out as early as 1906 by Pierre Duhem (1861–1916), a French physicist, mathematician, and philosopher of science, and later rediscovered by Willard Quine, the American philosopher. Duhem and Quine note that no hypothesis works as simply as $H_0 \Rightarrow O$. On the contrary, scientific hypotheses are accompanied by side conditions—instrumentation or controls—making the observation possible: H_0 and H_1 and H_2 and . . . $H_i \Rightarrow$ O. To believe a hypothesis that those specks changing places on successive nights are the moons of Jupiter you need to believe also that the telescope does what it purports to do when looking into the celestial sphere (Feyerabend 1975). To test the bending of starlight by the Sun you need to believe that the instruments are correctly calibrated. Dennis Lindley, who was in a position to know, writes that Harold Jeffreys "did not like Popper's views and tried to prevent his election to the Royal Society on the grounds that Popper could not do probability calculations correctly. Certainly Popper did make a serious error," Lindley says, "which illustrates how difficult probability calculations can be" (1991, 13).

The quite obvious richness of hypotheses H_0 and H_1 and H_2 and . . . H_i is the death knell of *modus tollens* and the simple hypothetico-deductive/Fisherian view of science. If one needs ignition *and* a supply of gasoline *and* a supply of oxygen, then a lack of combustion implies *either* a lack of ignition . . . *or* a lack of gasoline *or* a lack of oxygen *or* a lack of any number of other necessary conditions. So much for the simple "falsification" of H_0. In the case of a regression equation with many tacit variables (for IQ: nutrition, family, community, social class) or an experiment with many side conditions (no trucks rumbling down the street close to the laboratory and so forth), the falsification of a hypothesis, H_0, implies *either* that the hypothesis is wrong . . . *or* that one of the other variables or side conditions H_1, H_2, H_3, and so forth has intervened. *The Bell Curve,* by Richard J. Herrnstein and Charles Murray (1994), makes the usual Fisherian mistake: "Psychometrics approaches a table of correlations with one or another of its methods for factor analysis. . . . If they test traits in common, they are correlated, and if not, not. Factor analysis tells how many different underlying factors are necessary to account for the observed correlations between them" (581). No. As Lindley put it:

> It would be interesting to know how many significant results correspond to real differences [we reply: in economics less than 20 percent, in medicine and epidemiology between 10 and 30 percent, and in psychology less than 10 percent]. It is also interesting that many experimentalists, when asked what 5 percent significance means, often say that the probability of

the null hypothesis is 0.05. This is not true, save exceptionally. In saying this they are thinking like Jeffreys [or Gosset] but not acting like him. (1991, 12)

Precisely. *Modus tollens,* therefore, cannot be how science actually works, as the sons and daughters of Thomas Kuhn have been noting for decades. The children of Kuhn do not deny that a claimed falsification in the right circumstances is often persuasive. It is sometimes a sweet and useful argument. Both of us have used it from time to time in our scientific work. But falsification is nothing like the whole of scientific rhetoric. Near enough, the hypothetico-deductive model has been falsified. The sociologists and historians of science note that actual controversies in science are usually about whether this or that H_1, H_2, H_3 have intervened in the so-called crucial experiment. They note that the laws of science are metaphors and stories and are persuasive often for reasons other than merely their implied O's. And their O's are processed by actual scientists in ways that have more to do with Peirce's "abduction" than hypothetico-deductive deduction.

What is relevant here for the statistical case is that refutations of the null are trivially easy to achieve if power is low enough or the sample is large enough. The heroism of the Popperian tester of null hypotheses is not very impressive. Remember the six-inch hurdles. What falsificationism, strictly speaking, replaced was *confirmationism,* namely, that if you have a hypothesis H_0 that implies observations O, then if you observe O you can have more confidence in H_0. This is called in logic the fallacy of affirming the consequent. Because it was fallacious in simple logic—a logic that purposely *ignored* alternative hypotheses, power, and prior and posterior probabilities—the sons of the logical positivists such as Popper and the followers of Fisher sought a firmer ground in *modus tollens* for their deductive characterization of science.

But the so-called fallacy of affirming the consequent may not be a fallacy at all in a science that is serious about decisions and belief. It is after all how real scientists—such as Gosset, a lifelong Bayesian, Egon Pearson, a lifelong decision theorist and late in life a sympathizer with neo-Bayesianism, and Richard Feynman, a lifelong physicist and advocate of neo-Bayesianism—think. In his astonishing *Subjective and Objective Bayesian Statistics* the statistician James Press reports Feynman's view. Said Feynman, "The Bayesian [read: Jeffreys] approach is now the preferred method of comparing scientific theories . . . [so] to compare contending theories (in physics) one should use the Bayesian approach."[17]

Gosset and Feynman combine the confirmation approach with another, more commonly seen in bench scientists than in philosophers of science. It is what the physicist and philosopher of science Paul Feyerabend called "counterinduction": "A scientist who wishes to maximize the empirical content of the views he holds and who wants to understand them as clearly as he can must therefore introduce other views; that is, he must adopt a *pluralist methodology*."[18] And in the statistical terms relevant here, confirmationism and counterinduction are precisely the pluralist alternative to hard-boiled Fisherianism. Power, simulation, a variety of experiments, triangulation, actual replication, and exploratory data analysis leading to interocular trauma from the effect of magnitudes are different modes of affirming the consequent and are more generally a reasonable program of Gosset or Bayesian and Feynman confirmationism than is the dogma of Fisherian or Popperian falsificationism.

14

Medicine Seeks a Magic Pill

I once read a paper at a society showing that a widely believed hypothesis led to impossible consequences, and drew the conclusion that the hypothesis was wrong. A member said to me afterwards "I did not understand all that you were saying; at first you seemed to be arguing in favour of so-and-so and then to be arguing against it." The room was stuffy and he may have been asleep at a crucial moment; but when a logician of the standard of Bertrand Russell [the "member" in the story] said that we proved that a quantitative law could never acquire a high probability as a result of experiment I did think it was time to protest. What all this leads to is this: if a set of hypotheses, either possible scientific laws or axioms about the best ways of thought, leads to unacceptable consequences, then it is time to try to produce a different set that leads to acceptable ones.

HAROLD JEFFREYS 1963, 407

In general you wish to know the probability that your medical hypothesis, H, is true in view of the incomplete facts of the world. This is a problem of inference, inferring the likelihood of a result from data. If the symptoms of cholera start in the digestive system, then ingestion of something, perhaps foul water, is a probable cause. If cases of cholera in London in 1854 cluster around particular public wells, then foul water is probably a cause of cholera. The Fisher test does nothing to aid such judgments. It does the opposite. It measures the probability that the facts you are examining will occur, assuming the hypothesis H is true—the fallacy of the transposed conditional.

THE FALLACY OF THE TRANSPOSED CONDITIONAL

Suppose you are the sheriff arriving on the scene of *The Ox-Bow Incident*. Let the hypothesis H be "the character played by Anthony Quinn was hanged" (as in the film of 1943 he was) and the given fact—the data, O—be that "the character played by Anthony Quinn is dead." In general the probability of death by hanging is very low. People are not hanged every day in significant numbers. If, on the contrary, the given fact—the data, O—is that "Quinn was hanged" (we repeat that we mean the character played by him, not dear, dear Anthony) then the probability of the hypothesis "Quinn is dead" is very high—say, between 95 and 100 percent.

The Fisher test can shed light on the probability that "Quinn is dead" given that "Quinn was hanged." What the Fisher test wants to know and claims to measure is the opposite, the probability that Quinn was hanged, given that Quinn is dead. In societies unlike that of Ox-Bow, in which hanging is exceedingly rare, this probability, as we've noted, is close to zero. In fact even in 1885 Nevada the probability "Quinn was hanged" given that "Quinn is dead"—$Pr(H_0 \mid O)$—was low, radically lower anyway than a mistaken 5 percent philosophy would infer.

In a nonhanging society people die for many reasons other than hanging. People die from dehydration and cattle stampedes and bullet wounds. People die from cancer and flu and cardiac arrest. And therefore being dead is very weak evidence indeed that Quinn was hanged. Statistically speaking, the power of the test is not even defined. "Dead" is the datum, not the hypothesis. Being dead is "consistent with" the hypothesis that Quinn was hanged, to be sure, as the positivist rhetoric of the Fisherian argument emphasizes. But so what? A myriad of other hypotheses, very different from being hanged, and omitted from Fisherian models of unknown power and arbitrarily omitted variables, such as catching pneumonia or breaking your neck in a fall from your horse, are also consistent with it—"it" being the fact of being dead. The Fisherian test neither falsifies nor confirms. (In the film, incidentally, it was the lynch mob and not the sheriff that transposed hypothesis and data. They hung the wrong man.)

The multitalented Raymond Smullyan, a renowned logician, chess master, former Carnap student, magician, Tao mystic, comedian, and one-time teacher of piano at Roosevelt University, who was for many years a member of the faculty of philosophy at Indiana University and with whom one of us is acquainted, recounts an ancient joke about the Fisher fallacy (oomphful evidence of the joke's ancientness is just a moment away).

A student enters a professor's office, seeking advice on a course he might take. "Why not try a course in logic?" the professor replied. "What is logic?" "Logic," the professor explained, "is the study of how one factual statement can be deduced from another." "I don't get it," the student said, "Can you give me an example?" "Sure. Do you have lawn mower?" the professor asked. "Yes." "Then you have a lawn?" "Yes, I do have a lawn." "From which I deduce you have a house." "Yes, sir, I have a house. And I am married, too." "Do you and your wife have children?" "Yes, we do. We have two children." "From which I deduce that you are a heterosexual male."

"Gee," the student exclaimed, standing up, "this logic stuff is amazing! From the *fact* that I have a lawn mower, you can *deduce* that I am a heterosexual male!" and he dashed out of the professor's office and into a corridor, where he ran smack into a friend. "*Whoa!*" blurted the friend, "what's the rush?" "Well," the proud if breathless student replied, "*I've* just decided to enroll in a course on *logic!*" "What's logic?" the friend asked. "Logic is the study of how one factual statement can be deduced from another. For example, do you have a lawn mower?" "No," his friend said, "I do not have a lawn mower." To which the new student of logic replied, "You are *so* gay!"

Jacob Cohen made our point in a syllogism in his aptly titled article, "The Earth Is Round ($p < .05$)" (1994). "If a person is an American," Cohen writes, in a parody of the Fisherian logic, "then he is probably not a member of Congress. This person is a member of Congress. Therefore, he is probably not an American" (998). Cohen is pointing out that the illogic of being probably not an American is formally exactly the same as the Fisherian test of significance, which is the probabilistic equivalent of I do not have a lawn mower therefore I am not a heterosexual male. And it is mistaken. The structure of the logic is: hypothesize that $\Pr(O \mid H_0)$ is low; observe O in the data; conclude therefore that $\Pr(H_0 \mid O)$—*the transposed conditional of the original hypothesis*—is low. The argument appears implicitly or in less tasteful form in article after article in scientific journals. The fallacy is explicit in most statistics textbooks, smack in the center of the sections on hypothesis testing. It is wrong and for the same reason the mob in *The Ox-Bow Incident* was wrong.

Cohen examined real world consequences of the fallacious logic for an important topic in medicine and psychiatry, the diagnosis of adult-onset schizophrenia (1994, 999). In the United States the incidence of schizophrenia in adults is about 2 percent. Like hanging in 1885 Nevada, it is rare. Let H_0 = the person is normal (at any rate in the matter of schizo-phrenia), H_1 = the person is schizophrenic, and O = the test result on the person in question is positive for schizophrenia. A proposed screening test is estimated to have at least 95 percent accuracy in making the positive diagnosis (discovering schizophrenia) and about 97 percent accuracy in declaring a truly normal case "normal." Formally stated, Pr(normal | H_0) is approximately .97, and Pr(schizophrenic | H_1) > .95.

With a positive test for schizophrenia at hand, given the more than 95 percent assumed accuracy of the test, Pr(schizophrenic | H_0) is less than 5 percent—statistically significant, that is, at $p < .05$. In the face of such evidence a person in the Fisherian mode would reject the hypothesis of normal and conclude the person is schizophrenic.

But the probability of the hypothesis, given the data, is not what has been tested. The conditional probability has been transposed. The probability that the person is *normal*, given a positive test for schizophrenia, is in truth quite strong—about *60 percent*—not, as Fisherians believe, *less than* 3 percent, because, by Bayes's rule:

$$Pr(H_0 | O) = Pr(H_0) \times Pr(\text{test wrong} | H_0) / [Pr(H_0) \times Pr(\text{test wrong} | H_0)$$
$$+ Pr(H_1) \times (\text{Pr test right} | H_1)]$$
$$= (.98) \times (.03) / [(.98) \times (.03) + (.02)*(.95)] = .607,$$

a humanly important difference. Table 14.1 is Cohen's 2 × 2 table for a thousand cases. The table restates what Bayes's rule proves: the conditional probability of a case being normal though testing positively as schizophrenic is, Cohen exclaims, "not small—of the 50 cases testing as schizophrenic, 30 are false positives, actually normal, 60% of them!" (1994, 999).

The example shows how confused—and humanly and socially damaging—a fallacious conclusion from a 5 percent science can be. Think of the doctors standing around the Linda Blair character in the old thriller

TABLE 14.1. 5 Percent Statisticians Are Costly: The Transposed Conditional Overestimates Schizophrenia

Result	Normal	Schizophrenic	Total
Negative test (normal)	949	1	950
Positive test (schizophrenic)	30	20	50
Total	979	21	1,000

Source: Cohen 1994, 999–1000.

The Exorcist. One of the doctors (a Fisherian, it would seem) says of the possessed girl, "It's likely to be a case of split-personality disorder." Not likely, doctor. One of us has a good friend who as a child in the psychiatry-spooked 1950s was diagnosed as schizophrenic. The friend has evinced since then no symptom of the disease. But the erroneous diagnosis—an automatic result of the fallacy of the transposed conditional—has kept him ever since in a state of dull terror. Imagine in other arenas, with similarly realistically low priors, the damage done by the transposed conditional in diet pills or social welfare policy or commercial advertising or the foreign exchange markets. Once one seriously considers the concrete implications of such a large diagnostic error—such as believing that 3 percent of adults tested for schizophrenia are not schizophrenic when the truth is that *60 percent* of them are not schizophrenic—and realize that, after all, this magnitude of diagnostic error is running NASA and the departments of cardiovascular disease and breast cancer and HIV health policy, one should perhaps begin to worry about statistical significance.

Part of the problem historically was another campaign of Fisher's, following the elder Pearson: his killing off of Bayes's rule. Gosset, we've noted, was a lifelong Bayesian. He defended Bayesian methods against all comers: Karl Pearson, Egon Pearson, Jerzy Neyman, and Fisher.[1] Gosset in fact used Bayes's rule in his revolutionary articles of 1908 and crucially so in "The Probable Error of a Correlation Coefficient" (Student 1908b). In 1915 he wrote to the elder Pearson, "[I]f I didn't fear to waste your time I'd fight you on the *a priori* probability and give you choice of weapons! But I don't think the move is with me; I put my case on paper last time I wrote and doubt I've much to add to it" (quoted in E. Pearson 1990, 26–27). Gosset was courageous but mild in all his fights, including for Bayes's methods. In the warrior culture of hardboiled-dom he was not forceful enough.

We have discovered that Karl Pearson was in the classroom, at any rate according to his surviving lecture notes, more sensible than Fisher in his assessment of Bayes.[2]

> *Use of Bayes' Theorem—Give example*—Hence importance of using inverse probabilities. They are keynote to *inductive logic.* i.e. judging from experience to future events. Here we have exactly the same two points:
>
> (1). *A priori* ignorance of what will take place i.e. of its chance—
>
> (2). The experience of the past projected into the future as a basis for its prediction—i.e. permanence of statistical ratios. The permanence of

statistical ratios is bad on our experience of futures which become pasts. Our faith in permanence is also a probability band on a wide range of statistics.

Statistical ratios change in biological phenomena, do they change in physics? Growth of stability, oxygen, atomic weight of,—What is *an element?*—Chances in case of *a priori* ignorance.

(a.) Bayes hypothesis.—equal distribution of ignorance

(b.) Popular view, chance = .5 event as belief to fail as to succeed. Argument against this. Male & female births . . .

Pearson, you can see, was educating his students in Bayes's rule. But he wouldn't credit Bayes's rule in public.

Fisher was still more severe, as Sandy Zabell has shown, wholly intolerant of "inverse probability."[3] In Fisher's campaigns for maximum likelihood and his own notion of "fiducial probability" (one of the few campaigns of Fisher's that failed) he tried to kill off prior and posterior probability, and—at least with the mass of research workers as against the few highbrows—he largely succeeded. Egon Pearson and Neyman were at first persuaded by Fisher to turn from Bayes's rule.[4] But later in life, after Fisher died, Egon reverted to his original position. "Today in many circles," he said, "the current vogue is a neo-Bayesian one, which is of value because it calls attention to the fact that, in decision making, prior information must not be neglected."[5]

Medical scientists should be alarmed by the Fisherian notion that one would even consider "neglecting" prior information. Puzzled by what appeared to be a high variance in his heartbeat, one of us recently took a cardiac stress test, running on an increasingly steep and fast treadmill while connected to an electrocardiogram machine. While running along he explained in the way of professors to the female doctor present, a Northwestern University MD, the main point of our book. As the test and the brief lecture on the state of applied medical statistics ended, the doctor said, laughing, "You're a poster boy for cardiac health—*and* medically-substantive significance."

She had known and had worried about the infestation of *p*-values in medicine. But she had not known about the death of Bayes's rule in published medical research. In her daily duty at the hospital of rapidly assessing cardiovascular health, she noted, *of course* she depends on prior probability. "You don't have to wait for the logistic regression," she said, "to guess that a panting, obese, and cigarette-smoking fifty-five-year-old man who can't run the treadmill is in for it. *Can't.*" As the psychologist William Rozeboom said in 1960, "Insistence that published data must have the biases of the null-hypothesis-testing [or non-Bayesian] method built into the report, thus seducing the unwary reader into a perhaps highly inappropriate interpretation of the data, is a professional disservice of the first magnitude" (428). Clearly.

The 5 percenter longs to find a body of data "significant and consistent with" some hypothesis. The motive is by itself blameless. But, as Jeffreys observed long ago, the sequence of the 5 percenter's search procedure is backward and paradoxical (1963, 409). The 5 percenter is looking at the wrong thing in the wrong way. He's like a detective who interviews numerous suspects to a crime before the crime has been committed or like the delusional man on the street who waves on the coming traffic while standing in front of it. He assesses a null hypothesis relative to no concrete alternatives.

A bizarre example with medical implications can be found in the writings of G. Stanley Hall, a premier psychologist of the early twentieth century. Observing in 1904 a higher suicide rate among women (O), Hall strongly "accepts" his hypothesis (H_1) that women are evolutionarily inferior. Suicide, Hall writes, "is one expression of a profound psychic difference between the sexes."

> Woman's body and soul is phyletically older and more primitive, while man is more modern, variable, and less conservative. Women are always inclined to preserve old customs and ways of thinking. Women prefer passive methods; to give themselves up to the power of elemental forces, as gravity, when they throw themselves from heights or take poison, in which methods of suicide they surpass men. Havelock Ellis thinks drowning is becoming more frequent, and that therein women are becoming more womanly. (1904, 194, quoted in Gould 1981, 118)

One of us has an elderly aunt who can sit in the garden of a hot, Indiana summer evening untouched by mosquitoes. She chalks up her immunity to a side effect of a "nuclear treatment" received at midcentury to attack a tumor. Her sisters reply sarcastically, "Yes, dear, you're positively glowing!" Well, who's to deny her? Medical science since the arrival of Fisher's methods has had a problem with narrative. In the 1940s thoughtful people believed that the use of p's and t's in the design and evaluation of clinical trials would mark an advance over old wive's tales, crankery, anecdote, folkways, and fast-talking patent medicine salesmen. The dream of mechanization was as compelling in medicine as it was in war, social work, and the philosophy of mind. Case studies such as the Indiana aunt were projected onto the table of *t*. "Let the table decide." At 5 percent the medical scientists suddenly submitted, eyes locked hard in a sizeless stare. But the new method is just a mutation of old husband's tales, statistical crankery, probabilistic anecdote, scientific folkways, and fast-talking, twenty-first-century, statistical patent medicine salesmen.

IN EPIDEMIOLOGY, AS IN MEDICINE, THE TRANSPOSED CONDITIONAL ATTACHES TO THE SIZELESS STARE

In the 1994 volume of the *American Journal of Epidemiology* David A. Savitz, Kristi-Anne Tolo, and Charles Poole examined 246 articles published in the journal around the years 1970, 1980, and 1990. The articles were divided into three categories: infectious disease epidemiology, cancer epidemiology, and cardiovascular disease epidemiology. Each category contained for each date a minimum of 25 articles (see table 14.2). The main findings are presented in their figure 4, "Percent of articles published in the *American Journal of Epidemiology* classified as partially or completely reliant on statistical significance testing for the interpretation of the study results, by topic and time period" (1050). The findings are not surprising. The study shows that in 1990 some 60 to 70 percent of all cardiovascular and infectious disease epidemiologists relied exclusively on statistical significance as a criterion of epidemiological importance, as though fit were the same thing as importance. A larger share relied on the fallacy of the transposed conditional. The abuse was worse in 1990 than earlier.

The cancer researchers were less enchanted with statistical significance than cardiological and infectious disease researchers were, but they did

TABLE 14.2. The Savitz, Tolo, and Poole Study of "Significance" in the *American Journal of Epidemiology*

Fields	Years/Number of Articles		
	1970 Interval	1980 Interval	1990 Interval
Infectious disease	1970–71/27	1980/31	1990/29
Cancer	1967–74/25	1980–81/25	1990/31
Cardiovascular disease	1968–72/25	1980–81/27	1990–91/26

Source: Savitz, Tolo, and Poole 1994, table 1, 1049.

not reach standards of common sense. Savitz, Tolo, and Poole found that after a 60 percent reliance on a mere statistical significance in the early 1970s the abuse of p-values by cancer researchers actually fell. We don't know why. Maybe too many people had died. Still, 40 percent of all the cancer research articles in 1990 relied exclusively on Fisher's Rule of Two (1994, 1050).

In epidemiology, then, the sizeless stare of statistical significance is relatively recent, cancer research being an exception. In 1970 only about 20 percent of all articles on infectious disease epidemiology relied exclusively on tests of statistical significance. Confidence intervals and power calculations were of course absent. But epidemiology was not then an entirely statistical science. Only about 40 percent of all empirical articles in infectious disease epidemiology employed some kind of statistical test. But suddenly significance took hold, and by 1980 some 40 percent relied exclusively on the tests (compare our "question 16" in economics, where in the 1980s it was about 70 percent). And by 1990 most subfields of epidemiology had, like economics and psychology, become predominately Fisherian. Statistical significance came to mean "epidemiological significance." Statistical *insignificance* came to mean "ignore the results." Remember Vioxx.

Medicine Doesn't Pass the Test

Douglas G. Altman, a statistician and cancer researcher at the Medical Statistics Laboratory in London, has been watching the use of medical statistics, and especially the deployment of significance testing, for twenty years. In 1991 Altman published an article called "Statistics in Medical Journals: Developments in the 1980s." The article appeared in *Statistics in Medicine*. Altman's experience had been similar to ours in economics. At conferences and seminars Altman's colleagues would claim to be convinced that the abuse of t-testing had by the 1980s abated and was practiced only by the less competent medical scientists. Any thoughtful reader of the journals knew that such claims were false. To bias the results in favor of the defenders of the status quo Altman examined the first one hundred "original articles" published in the 1980s in the *New England Journal of Medicine*. These were new and full-length research articles based on never before released or published data from clinical studies or other methods of observation. Altman's sample design was meant to replicate for comparative purposes an earlier study by Emerson and Colditz (1983), who studied the matter in 1978–79 (Altman 1991, 1899).

The findings:

It is my impression that the trends noted by Felson et al. have continued throughout the 1980s. . . . The obsession with significant p values is seen in several other ways:

(i) Reporting of [statistically] significant results rather than those of most importance (especially in abstracts);

(ii) The use of hypothesis tests when none is appropriate (such as for comparing two methods of measurements or two observers);

(iii) The automatic equating of statistically significant with clinically important, and non-significant with non-existent;

(iv) The designation of studies that do or do not "achieve" significance as "positive" or "negative" respectively, and the common associated phrase "failed to reach statistical significance." . . . A review [by other investigators] of 142 articles in three general medical journals found that in almost all cases (1076/1092) researchers' interpretations of the "quantitative"(that is, clinical) significance of their results agreed with statistical significance. Thus across all medical areas and sample size p rules, and $p < 0.05$ rules most. It is not surprising if some editors share these attitudes, as most will have passed through the same research phase of their careers and some are still active researchers. (Altman 1991, 1906)

Altman was not surprised when he found in medicine, as we were not surprised in economics, that his colleagues were deluding themselves. "I noted in the first issue of *Statistics in Medicine* that most journals gave much more attention to the format of references in submitted articles than they gave to the statistical content," Altman wrote. "This remains true" (1991, 1900). Editors are much exercised, he observed with gentle sarcasm, over whether to use "P, p, *P*, or *p* values" (1902)—but pay no heed to oomph. "It is impossibly idealistic," Altman believed, "to hope that we can stop the misuse of statistics, but we can apply a tourniquet . . . by continuing to press journals to improve their ways" (1908).

Steven Goodman, in a meaty piece on the "p-value fallacy" published in the *Annals of Internal Medicine,* observed ruefully that "biological understanding and previous research play little formal role in the interpretation of quantitative results." That is, Bayes's rule is set aside, as is the total quality management of medical science, the seeing of results in their context of biological common sense. "This [narrowly Fisherian] statistical approach," Goodman writes, "the key components of which are P values and hypothesis tests, is widely perceived as a mathematically coherent approach to inference. There is little appreciation in the medical community that the methodology is an amalgam of incompatible elements."[6]

Altman, Savitz, Goodman, and company are not rare dismissible madmen, high on the wrong medicine. According to Altman, between 1966 and 1986 fully *150* articles were published criticizing the use of statistics in medical research (1991, 1897). 150. The studies agreed that Fisher significance in medical science had become the nearly exclusive technique for making a quantitative decision and that statistical significance had become in the minds of medical writers equated increasingly, and erroneously, with clinical significance. You yourself may be killed by it.

15

Rothman's Revolt

It is curious how often the most acute and powerful intellects have gone astray in the calculation of probabilities.
WILLIAM STANLEY JEVONS 1877, 231

Declarations of "significance" or its absence can supplant the need for any real interpretation of data; the declarations can serve as a mechanical substitute for thought, promulgated by the inertia of training and common practice.
KENNETH J. ROTHMAN 1986, 118

As early as 1978 the situation was sufficiently dire that two contributors to the *New England Journal of Medicine*, Drummond Rennie and Kenneth J. Rothman, published op-ed pieces in the journal pages about the matter (Rennie 1978; Rothman 1978). Rennie, the deputy editor of the journal—and in the 2000s the deputy editor of the *Journal of the American Medical Association*—was not critical of his colleagues' practice. But Rothman, who was a young associate professor at Harvard and the youngest member of the editorial board, blasted away. In "A Show of Confidence" he made a crushing case for measuring *clinical* significance, not statistical significance. Citing a Freiman et al. article on "71 Negative [Clinical] Trials" (1978) Rothman argued that the measurement and interpretation of size of effects, confidence intervals, and examination of power functions with respect to effect size (à la Freiman et al. by graphic demonstration) was the better way forward.

DEAR AUTHOR: ABOUT THOSE *P*-VALUES

Rothman—an epidemiologist and biostatistician with a lifelong interest in the rhetoric of his fields—wanted to shift the statistical rhetoric of his fields toward the promised land of Gosset and Jeffreys. Rennie and the other editors decided on a different solution. Original articles would be subjected to a prepublication screening by a professional statistician. Rothman was at first hopeful, thinking statistical review would repair the journal. The director of statistical reviews was well chosen—the late Frederick Mosteller (1916–2006), the founder of Harvard's Statistics Department and a giant of twentieth-century data analysis. But Mosteller was only the director, not the worker. Rothman tells us that he as the inside critic and Mosteller as the outside director had not been able to do anything together to raise the standards.[1] The problem with prepublication statistical review, of course, is that the articles go not to the Rothmans and Mostellers and Kruskals but out to Promising Young Jones in the outer suburbs dazzled by his recently mastered 5 percent textbooks. An example nowadays is the "Statistical Analysis Plan," aptly acronymized as SAP, which lays down the minimum statistical criteria considered acceptable by the Food and Drug Administration.

Ironically, Harold Hotelling had worried over the problem in 1940, writing that "every university department has a bright graduate student whose placement is an immediate problem."

> Young Jones has already demonstrated a quantitative turn of mind in the course on Money and Banking, or in the Ph.D. thesis on which he has already made substantial progress, dealing with The Proportion of Public School Yard Areas Surfaced with Gravel. He may even recall having had a high-school course in trigonometry. His personality is all that might be desired. . . . And so the "Instructor [or prepublication reviewer] to be announced" materializes as Jones. (1940, 14)

Rothman complained in his editorial in the *New England Journal* that Fisherian "testing . . . is equivalent to funneling all interest into the precise location of one boundary of a confidence interval" (1978, 1363). It continued to funnel furiously in the pages of *Lancet,* the *British Medical Journal,* the *New England Journal of Medicine,* and hundreds of other journals of medicine throughout the 1980s.

Rothman then became assistant editor of the *American Journal of Public Health* (*AJPH*). The chief editor "seemed to be sympathetic" with Rothman's views—Rothman recalls one time when the editor backed him up in a little feud with a well-placed statistician. Still, Rothman's views hardly set journal policy, and it shows in the journal. Rothman finally found his chance when in 1990, after fifteen years of quiet struggle, he started his own journal, *Epidemiology.*

His editorial letter to potential authors was unprecedented.

> When writing for *Epidemiology,* you can . . . enhance your prospects if you omit tests of statistical significance. . . . In *Epidemiology,* we do not publish them at all. . . . We discourage the use of this type of thinking in the data analysis, such as in the use of stepwise regression. We also would like to see the interpretation of a study based not on statistical significance, or lack of it, for one or more study variables, but rather on careful quantitative consideration of the data in light of competing explanations for the findings. For example, we prefer a researcher to consider whether the magnitude of an estimated effect could be readily explained by uncontrolled confounding or selection biases, rather than simply to offer the uninspired interpretation that the estimated effect is "significant." . . . Misleading signals occur when a trivial effect is found to be "significant," as often happens in large studies, or when a strong relation is found "nonsignificant," as often happens in small studies. (Rothman 1998, 334)

Like Gosset and Jeffreys and Zellner, Rothman doubted the philosophical grounding of p-values (334). As Jeffreys put it:

> If P is small, that means that there have been unexpectedly large departures from prediction [under the null hypothesis]. But why should these be stated in terms of P? The latter gives the probability of departures, measured in a particular way, equal to *or greater than* the observed set, and the contribution from the actual value [of the test statistic] is nearly always negligible. *What the use of* P *implies, therefore, is that a hypothesis that may be true may be rejected because it has not predicted observable results that have not occurred.* This seems a remarkable procedure. On the face of it the fact that such results have not occurred might more reasonably be taken as evidence for the law [or null hypothesis], not against it. The same applies to all the current significance tests based on P integrals. (Jeffreys 1961, 385; editorial insertions by Arnold Zellner [1984, 288]; italics in original)

Rothman concluded the letter by offering advice on how to publish, quantitatively, epidemiologically significant figures such as odds ratios on specific medical risks bounded by confidence intervals.

Now with his own journal, Rothman was going to get it right. In January 1990 he and the associate editors Janet Lang and Cristina Cann published another luminous editorial, "That Confounded *P*-Value" (Lang, Rothman, and Cann 1998). They "reluctantly" (8) agreed to publish *p*-values when "no other" alternative was at hand. But they strongly suggested that authors of submitted manuscripts illustrate "size of effect" (7) in "figures"—in plots of effect size.

Rothman and his associates were and are not alone, even in epidemiology. The distinguished statistician James O. Berger (2003) has recently shown how epidemiologists and other sizeless scientists go wrong with *p*-values. Use of Berger's *applet,* a public-access program, shows Rothman's skepticism to be empirically sound.[2] The program simulates a series of tests, recording how often a null hypothesis is "true" in a range of different *p*-values. Berger cites a 2001 study by the epidemiologists J. A. C. Sterne and G. Davey Smith, which found that "roughly 90% of the null hypotheses in the epidemiology literature are initially true." Berger reports that even when *p* "is near 0.05, at least 72%—and typically over 90%" of the null hypotheses will be true.[3] Berger agrees with Rothman and the authors here that, on the contrary, "true" is a matter of judgment—a judgment of *epidemiological,* not mere statistical, significance. It is about the quality of the water from the London wells.

Rothman's letter is only the most explicit of a thin, bright stream of such declarations from editors in the statistical sciences. In our own field we have Morris Altman's in the *Journal of Socio-Economics,* Diana Strassman's in *Feminist Economics,* and Marc Gaudry's in his journal of transportation economics. And in psychology, as we have noted, there have been a couple of dozen such editorial cries of anguish.[4]

Rothman's letter itself elicited no response. This is our experience, too: many of the Fisherians, to put it bluntly, seem to be less than courageous in defending their views. Hardly ever have we seen or heard an attempt to provide a coherent—or indeed any—response to the case against null-hypothesis testing for "significance." The only published response that Rothman can recollect in epidemiology came years before the letter from J. L. Fleiss, a prominent biostatistician, in the *American Journal of Public Health* (1986). But Fleiss merely complained that "an insidious message is being sent to researchers in epidemiology that tests of significance are invalid and have no place in their research" (559). He gave no actual *arguments* for giving Fisherian practices a place in research. This is similar to our experience. Kevin Hoover and Mark Siegler offered in 2005 (published in 2008) the only written response to our complaints in economics that we have seen. Courageous though it was for them to venture out in defense of the Fisherian conventions, a sterling exception to the nervous silence of their colleagues, they could offer no actual arguments (although, as we have noted, they did catch us in a most embarrassing failure to take all the data, as we thought we had, from the *American Economic Review* in the 1990s). Hoover and Siegler merely wax wroth for many pages against our strictures.

On the *Truck and Barter* blog Kevin Brancato of the Rand Corporation looked into the matter with some care, starting from a position sympathetic with Hoover and Siegler. But of their blast he concluded, "I must say that I'm disappointed. . . . I don't think Hoover and Siegler have much new to say other than [that] the problem is not as bad as Ziliak and McCloskey claim. However, this is an empirical question, that in my mind, Hoover and Siegler fail to address thoroughly—in fact, not even in a cursory fashion. . . . In sum: A majority of papers in the *AER* in the 1980's and 1990's did not distinguish economic and statistical significance, although trends in the share are not yet determinable." And a bit later the distinguished economist Thomas Mayer, who has for a long time been making the same point about econometric practice as we make here, remarked: "I don't understand the relevance of the table with the additional papers. Isn't the main issue whether authors pay attention to the size of the coefficients? And that is not in the table. What am I missing?" Professor Mayer is not missing anything: the main issue is, as Mayer suggests, not whether the Ziliak and McCloskey survey as originally published contained "the entire population" but whether, on the evidence of any reasonable sample of statistical practice in the *American Economic Review* (100 percent or an as-it-happens random sample of less than 100 percent), the authors pay attention to the size of the coefficients. A large majority—over 80 percent—do not. About *this* error, our critics Hoover and Siegler, and it appears the rest of the econometric profession, have nothing to say. "Not even," as Brancato points out, "in a cursory fashion."

Even the rare courageous Fisherians, in other words, do not deign to make a case for their procedures. They merely complain that the procedures are being criticized. "Other defenses of [null-hypothesis significance testing]," Fidler et al. observed, "are hard to find."[5] The Fisherians, being comfortably in control, appear inclined to leave things as they are, sans argument. One can understand. If you don't have any arguments for an intellectual habit of a lifetime perhaps it is best to keep quiet.

A REVOLT THAT FAILED?
Epidemiology AND THE FIDLER STUDY

Rothman's campaign did not succeed, as an article in 2004 jointly authored by Fidler, Thomason, Cumming, Finch, and Leeman concluded (Fidler et al. 2004b). They found, as we and others have found in economics and psychology and other fields of medicine, that epidemiology is getting worse despite Rothman. Over 88 percent of more than seven hundred articles they reviewed in *Epidemiology* (between 1990 and 2000) and the *American Journal of Public Health* (between 1982 and 2000) failed, they find, to distinguish and interpret substantive significance. In the *American Journal of Public Health* some 90 percent confused a statistically significant result with an epidemiologically significant result and equated statistical *insignificance* with substantive unimportance. Epidemiology journals, in other words, performed worse than the *New England Journal of Medicine,* Rothman's training ground as editor.

Fidler and her coauthors observe that for decades "advocates of statistical reform in psychology have recommended confidence intervals as an alternative (or at least a supplement) to *p* values" (Fidler et al. 2004b, 120). The American Psychological Association's *Publication Manual* called confidence intervals in 2001 "the best reporting strategy," though few seem to be paying attention.[6] Since the mid-1980s confidence intervals have been widely reported in medical journals. Unhappily, requiring the calculation of confidence intervals, contrary to our optimistic hope, does not guarantee that effect sizes will be interpreted more carefully or indeed at all. Savitz, Tolo, and Poole (1994) find that even though 70 percent of articles in the *American Journal of Epidemiology* report confidence intervals "inferences are made regarding statistical significance tests, often based on the location of the null value with[out] respect to the bounds of the confidence interval" (1051). In other words, say Fidler and her coauthors, confidence intervals "were simply used to do [the null-hypothesis testing ritual]" (2004b, 120).

They also point out that there has been "little interdisciplinary discussion of null-hypothesis significance testing" (Fidler et al. 2004b, 119). The fact is strange but true. The recent "perestroika" in political science has forced rational-choice theorists and other econowannabes to pause in their imposition of Max U and *t*-tests on the field. But pause only, we've noted, not reverse their descent to substantive insignificance. The young economists in the so called post-autistic movement, many members of the Asso-

ciation for Social Economics, scores of Marxists, and a few other economists, especially those associated with feminist economics and Daniel
Klein's *Econ Journal Watch* (such as the two of us), have applauded moves
toward substance. Long before, in their 1970 anthology, *The Significance
Test Controversy* (1970), the sociologists Denton Morrison and Ramon
Henkel brought together psychologists and sociologists (and one economist) to conspire against the hegemony of Fisher. But even these sizable
steps toward an interdisciplinary recognition of the problem failed.

Fidler and her coauthors attempted, as we have, to assemble outside
allies. They "sought lessons for psychology from medicine's experience
with statistical reform by investigating two attempts by Kenneth Rothman to change statistical practices." They examined 594 *American Journal of Public Health* articles published between 1982 and 2000 and 110
Epidemiology articles published in 1990 and 2000.

> Rothman's editorial instruction to report confidence intervals and not p
> values was largely effective: In *AJPH*, sole reliance on p values dropped
> from 63% to 5%, and confidence interval reporting rose from 10% to
> 54%; *Epidemiology* showed even stronger compliance. However, com-
> pliance was superficial: Very few authors referred to confidence intervals
> when discussing results. The results of our survey support what other
> research has indicated: Editorial policy alone is not a sufficient mecha-
> nism for statistical reform. (Fidler et al. 2004b, 119)

Rothman himself has said of his attempt to reduce p-value reporting in his
Epidemiology that "my revise-and-resubmit letters . . . were not a covert
attempt to engineer a new policy, but simply my attempt to do my job as
I understood it. Just as I corrected grammatical errors, I corrected what I
saw as conceptual errors in describing data."[7]

Fidler's team studied the *American Journal of Public Health* and *Epi
demiology* before, during, and after Rothman's editorial stints, before and
after the International Committee of Medical Journal Editors' creation of
statistical regulations encouraging the analysis of effect size, and before
and after the changes to the *AJPH*'s "Instructions to Authors" encouraging the use of confidence intervals. Rothman as assistant editor of course
did not make policy at the journal. He made his own preferences known
to authors, but ultimately he "carried out the editor's policy," which only
occasionally overlapped with Rothman's ideal.[8]

Fidler et al. counted a statistical practice "present," such as asterisk biometrics, the ranking of coefficients according to the size of the p-value, if

an article contained at least one instance of it. Their full questionnaire is similar to ours in economics, focusing on substantive as against statistical significance testing. Did *significant* mean "epidemiologically important" or "statistically significant"? Practice was recorded as ambiguous if the author or authors did not preface *significant* with *statistically*, follow the statement of significance directly with a *p*-value or test statistic, or otherwise differentiate between statistical and substantive interpretations. *Explicit power* in their checklist means "did a power calculation." *Implicit power* means some mention of a relationship between sample size, effect size, and statistical significance was made—for example, a reference to small sample size as perhaps explaining failure to find statistical significance.

The results, alas:

> Of the 594 *AJPH* articles, 273 (46%) reported null hypothesis significance tests. In almost two thirds of the cases "significant" was used ambiguously. Only 3% calculated power and 15% reported "implied power." [A]n overwhelming 82% of null-hypothesis-siginificance-test articles had neither an explicit nor implicit reference to statistical power, even though all reported at least one non-significant result. (Fidler et al. 2004b).

Fifty-four percent of *American Journal of Public Health* articles reported confidence intervals; 86 percent did in *Epidemiology*. But "Table 2 shows that fewer than 12 percent of *AJPH* articles with confidence intervals interpreted them and that, despite fully 86 percent of articles in *Epidemiology* reporting confidence intervals, interpretation was just as rare in that journal."[9] The situation, they find, did not improve with the years. The authors usually did not refer in their texts to the width of their confidence intervals and did not discuss what is epidemiologically or biologically or socially or clinically significant in the size of the effect. In other words, during the past two decades more than six hundred of some seven hundred articles published in the leading journals of public health and epidemiology showed no concern with epidemiological significance. Thus, too, economics, sociology, population biology, and other Fisherian fields.

Our reading diverges from that of Fidler et al. in only one respect. Rothman, we find, was in his own journal quite successful in getting authors to focus on the epidemiological meaning of their empirical results. Interpretation of effect size was the primary concern of nearly every article we examined, and only the rare author seemed to believe that statistical significance at conventional levels was necessary for proving a scientific result. And, contrary to the depressing claim of Fidler et al., none of the authors in *Epidemiology* we read seemed to believe that statistical significance sufficed. Of sixteen "original articles" published in the premier issue (January 1990), for example, six were both quantitative and nonmethodological. Of those six articles, three used tests of statistical significance. Each of the three articles reported confidence-interval widths, and each of them focused on effect size, usually in terms of the epidemiological meaning of estimated odds ratios. Two of the methodological papers (including one by Rothman himself) were sharply critical of conventional significance testing and suggested alternative methods. Of seventeen original articles published in January 1996 (vol. 7, no. 1), fifteen were both quantitative and nonmethodological. Of those fifteen articles, thirteen used tests of statistical significance. All reported confidence intervals, and fourteen of fifteen showed the widths of the confidence intervals *and* interpreted effect sizes. One article (Mink et al. 1996), a study concerning cigarette smoking and ovarian cancer, relied on the conventional qualitative standard of statistical significance, and another (Schwartz 1996, 23–24) seemed a little overcharmed by statistical fit. But only Mink et al. relied heavily on significance testing and, despite it, did finally interpret the epidemiological meaning of risk ratios. Schwartz did not rely entirely on fit and focused on the interpretation of effect size (23, 25). The pattern in January 2000 (vol. 11, no. 1), the last issue of the journal before Rothman and his associates passed the torch to Allen Wilcox and others, is about the same.

The standard of clinical or epidemiological significance in *Epidemiology*, January 1990 to January 2001, is in our estimation quite high—truly exemplary among the testimating sciences. Rothman's journal resembles the best of radiology or cell biology or our own field of economic history. The arguments focus on quantitative standards established in previous empirical studies and almost entirely avoid the sizeless stare. Anne Prener, G. Enghom, and O. M. Jensen, in "Genital Anomalies and Risk for Testicular Cancer in Danish Men," is typical.

> Adjustments [of crude data] in the logistic regression analysis for birth order, social class, birth weight, and year of birth increased the RR [relative risk ratio] to 5.2 (95% confidence interval = 3.1–13.0). Men with a history of undescended testis were 7.3 times more likely to develop testicular seminoma, whereas undescended testis was associated with an RR of 3.6 for nonseminoma testicular cancer. (1996, 15)

In the concluding remarks, Prener and her coauthors compare the risk for testicular cancer to risks found in related studies, and the confidence bands are always in sight (17). Prener's is a meaningful quantitative rhetoric.

EDITOR EFFECT

Still, Fidler et al. are in a larger sense correct. *Most* epidemiological research is bamboozled by statistical significance. When in 2000 Rothman left his post as editor of *Epidemiology* confidence-interval reporting remained high—it had become common in medical journals. But in the *American Journal of Public Health* reporting of unqualified *p* "again became common." Rothman's success at *Epidemiology* appears to have been longer lasting. Still, interpretation in other journals of epidemiology is rare. "In both journals [Fidler et al. should add 'but not in *Epidemiology*'] . . . when confidence intervals were reported, they were rarely used to interpret results or comment on [substantive] precision. This rather ominous finding holds even for the most recent years we surveyed" (Fidler et al. 2004b, 123). Fidler and her team confirm in thousands of tests what Savitz, Tolo, and Poole found in the *American Journal of Epidemiology* in tens of thousands of tests (1994) and what Rossi found in 39,863 tests in psychology and speech and education and sociology and management (1990, 648). See Table 15.1.

It appears that one editor working in isolation cannot turn a science equipped with personal computers and canned programs away from the Significance Mistake, even well-placed editors speaking from important scientific institutions. A Rothman or Altman can affect the price of sig-

TABLE 15.1. The Fidler Sample

Publication Year	Number of Articles Coded	Reason for Choosing Year
1982	67	Pre-Rothman
1986	98	Expected maximum influence of Rothman, whose term was 1984 to February 1987
1988	71	Immediate post-Rothman
1989	72	Post-Rothman
1990	72	Post-Rothman and post-ICMJE[a] recommendations (published 1988, referred to in *AJPH*[b] "instructions to authors" in 1989)
1993	72	New editor and specific reference to ICMJE recommendations dropped from "instructions to authors" in 1991
1994	72	As for 1993
2000	70	Recent practices

Source: Fidler et al. 2004b, 121, table 1.
[a]International Committee of Medical Journal Editors.
[b]*American Journal of Public Health.*

nificance in his own journal. But in the larger market, as economists say, he is a price taker.

The historian of medicine Richard Shyrock argued long ago that instruments such as the stethoscope and the X-ray machine saved some parts of medicine from the Fisherian pitfall. If one can see or hear the problem, one does not need to rely on correlations (1961, 228). Since 1961, though, doctors have lost many of their skills of physical assessment, even with the stethoscope (and certainly with their hands) and have come to rely on a medical literature deeply infected with Fisherianism. Shyrock's piece appeared in 1961 in a special issue of *Isis* on the history of quantification in the sciences, mostly celebrating the statistical side of it. Puzzlingly, none of the contributors to the symposium mentioned the Gosset-Fisher-Neyman-Pearson-Jeffreys-Deming-Savage complex. Fisher-significance, the omission suggests, was not to be put on trial. The inference machines remained broken.

16

On Drugs, Disability, and Death

Whether statisticians like it or not, their results are used to decide between hypotheses, and it is elementary that if p *entails* q, q *does not necessarily entail* p. *We cannot get from "the data are unlikely given the hypothesis" to "the hypothesis is unlikely given the data" without some additional rule of thought. Those that reject inverse probability have to replace it by some circumlocution, which leaves it to the student to spot where the change of data has been slipped in[, in] the hope that it will not be noticed.*

HAROLD JEFFREYS 1963, 409

By 1988 the International Committee of Medical Journal Editors had been sufficiently pressured by the Rothmans and Altmans to revise their "uniform requirements for manuscripts submitted to biomedical journals." "When possible," the committee wrote, "quantify findings and present them with appropriate indicators of measurement error or uncertainty (such as confidence intervals). Avoid sole reliance on statistical hypothesis testing, such as the use of *p* values, which fail to convey important quantitative information"(in Fidler et al. 2004b, 120). The formulation is not ideal. The "error" in question is tacitly understood to be sampling error alone when after all a good deal of error does not arise from the smallness of samples. "Avoid sole reliance" on the significance error should be "Don't commit" the significance error. The "important quantitative information" is effect size, which should have been mentioned explicitly. Still, it was a good first step and in 1988 among the sizeless sciences was amazing.

The requirements—on which at a formative stage Rothman, among others, had contributed an opinion—were widely published. They appeared, for instance, in the *Annals of Internal Medicine*—where later the Vioxx study was published—and in the *British Medical Journal*. More than three hundred medical and biomedical journals, including the *American Journal of Public Health*, notified the international committee of their willingness to comply with the manuscript guidelines (Fidler et al. 2004b, 120). But the requirements have not helped.

"Significant" Temptations to Use Drugs

The essence of the problem of reform—and the proof that we need to change academic and institutional incentives, including criteria for winning grants—is well illustrated in a study we have already mentioned of "temptation to use drugs" published in the *Journal of Drug Issues*. The study was financed by the Centers for Disease Control. It was authored by two professors of public health at Emory University (one of them an associate dean for research) and a third professor, a medical sociologist at Georgia State University. The study was conducted in Atlanta between August 1997 and August 2000. Its subjects were African American women—mothers and their daughters—living in low-income neighborhoods of Atlanta.[1] The dependent variable was "frequency-of-[drug] use and times-per-day" multiplied for each drug type and summed by month. In the 125 women studied the value of the dependent variable ranged from zero to 910, that is, from zero to an appalling thirty drug doses a day. Statistical significance decides everything.

> Initially, each of the temptations-to-use drugs variables was entered into simple regression equations, to determine if they were *statistically significant* predictors of the outcome measure. Next, those found to be *related to amount of drug use* reported were entered simultaneously into a stepwise multiple regression equation. . . . Next, the bivariate relationships between the other predictor variables listed earlier were examined one by one, using Student's *t* tests whenever the independent variable was dichotomous. . . . Items that were found to be marginally- or statistically-significant predictors in these bivariate analyses were selected for entry into the multivariate equation. (Klein, Elifson, and Sterk 2003, 169, 170)

The authors do at least report mean values of the temptations to use drugs—a first step in determining substantive significance. For example, they report that women were "least tempted to use drugs when they were talking and relaxing (74.0%), experiencing withdrawal symptoms

(73.3%), [and] waking up and facing a difficult day (70.7%). And they would be tempted "quite a bit" or "a lot" when they were "with a partner or close friend who was using drugs (38.5%)" or when "seeing another person using and enjoying drugs (36.1%)" [170]. Recall how the authors presented their findings.

> while with friends at a party ($p < .001$), while talking and relaxing ($p < .001$), while with a partner or close friend who is using drugs ($p < .001$), while hanging around the neighborhood ($p < .001$), when happy and celebrating ($p < .001$), when seeing someone using and enjoying drugs ($p < .05$), when waking up and facing a tough day ($p < .001$),

And on and on. Remember that the article concluded that "The only item that was not associated with the amount of drugs women used was 'when one realized that stopping drugs was extremely difficult'" (172).

This is a hoax, perhaps a belated retaliation for the 1990s *Social Text* scandal, in which a scientist posed as a postmodern theorist to expose intellectual pretense. Alas, it's not. What is the scientific or policy oomph of such a temptations-to-use-drugs study? Everything is alleged to be "significant."

Be Not Tempted by Other False Idols, Such as "False Negatives"

In September 1978 Jennie A. Freiman, Thomas C. Chalmers, Harry Smith Jr., and Roy R. Kuebler, doctors and statistical researchers at Mount Sinai in New York, published in the *New England Journal of Medicine* a study entitled "The Importance of Beta, the Type II Error, and Sample Size in the Design and Interpretation of the Randomized Control Trial."

The abstract reads:

> Seventy-one "negative" randomized control trials were re-examined to determine if the investigators had studied large enough samples to give a high probability (>0.90) of detecting a 25 per cent and 50 per cent therapeutic improvement in the response. *Sixty-seven of the trials had a greater than 10 per cent risk of missing a true 25 per cent therapeutic improvement, and with the same risk, 50 of the trials could have missed a 50 per cent improvement.* Estimates of 90 per cent confidence intervals for the true improvement in each trial showed that in 57 of these "negative" trials, a potential 25 per cent improvement was possible, and 34 of the trials showed a potential 50 per cent improvement. Many of the

therapies labeled as "no different from control" in trials using inadequate samples have not received a fair test. Concern for the probability of missing an important therapeutic improvement because of small sample sizes deserves more attention in the planning of clinical trials. (Freiman et al. 1978, 690; italics supplied)

Freiman, who is a specialist in obstetrics and gynecology, and her colleagues, in other words, had reanalyzed seventy-one articles in medical journals. Heart- and cancer-related treatments dominated the clinical trials under review. Each of the seventy-one articles concluded that the "treatment"—for example, "chemotherapy" or "an aspirin pill"—performed no better in a clinical sense than did the "control" of nontreatment or a placebo. That is, the treatments were "insignificant."

Freiman et al. found that if the authors of the original studies had considered the power of their tests—the probability of rejecting the null hypothesis "[treatment] no different from control" as the treatment effect moves in the direction of "vast improvement"—meaning in conjunction with effect size, the experiments would not have ended "negatively." That is, the clinicians conducting the original studies would have found that indeed the treatment therapy such as aspirin was capable of producing "important therapeutic improvement."

Specifically, Freiman et al. found that if fully fifty of the seventy-one trials had paid attention to power and effect size, and not merely to a one-sided, qualitative, yes/no interpretation of "significance," they would have *reversed* their conclusions. Astonishingly, they would have found up to "50 per cent improvement" in "therapeutic effect." The point here about hearts and cancer is the same as the Gosset point about Guinness beer, which is the same as the Neyman and Pearson point about justice, which is the same as the Jeffreys point about *p*-values, which is the same as the McCloskey point about purchasing power parity, which is the same as the Ziliak point about black unemployment rates. The Fisherian tests of significance, the only tests employed by the original authors of the seventy-one studies, literally could not see the beneficial effects of the therapies under study, though staring at them.

Freiman and her team were being conservative. Consider figure 16.1. The vertical axis is β, a measure of Type II error, that is, the probability of accepting the hypothesis "treatment has no good effect" when in fact the treatment has good effect. (One minus this β, then, is the "power" of the test at various effect sizes.) The horizontal axis shows the treatment effect size, defined by "reduction in mortality." The downward-sloping

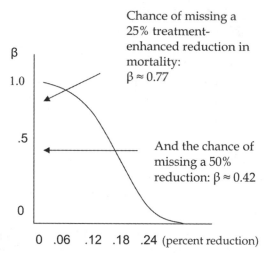

Fig. 16.1. Neglecting power is a bad idea for humans. Percent reduction in mortality from a baseline level of 29.7 percent (origin). $(1 - \beta)$ is the power of the test: for many phenomena, including medical, power is an increasing function of effect size. (Adapted from Freiman et al. 1978, fig. 1, 691. Copyright © 1978 Massachusetts Medical Society.)

curve inside the two axes shows how the Type II error (the error of gullibility) falls as the treatment effect (the reduction in mortality) increases. That is, it shows how power increases as effect sizes depart further and further from the null.

The precise standard of improvement—the minimum standard of oomph the authors set—is a "reduction in mortality from the control [group] mortality rate," a baseline rate of 29.7 percent (Freiman et al. 1978, 691). They realize it is not a very strict standard of medical oomph. They are bending over backward not to find their colleagues mistaken. Like Gosset, they want to give their Fisherian colleagues the benefit of the doubt.

Yet they found that 70 percent of the alleged "negative" trials were prematurely stopped, missing an opportunity to reduce the mortality of their patients by up to 50 percent. Of the patients who were prescribed sugar pills or otherwise dismissed, in other words, about 30 percent died unnecessarily. In one typical article the authors in fact missed at $\alpha = 0.05$ a 25 percent reduction in mortality with probability about 0.77 and, at the same level of Type I error, a 50 percent reduction with probability about 0.42 (Freiman et al. 1978, 691).

Each of the seventy-one experiments was shut down on the belief that a 30 percent death rate was equally likely with the sugar pill (or whatever the control was) and with the treatment therapy, spurning opportunities to save lives. The article shows that in the original experiments as few as 15 percent of the patients receiving the treatment therapy would have died had the experiment continued—half as many as actually died.

We agree with Rothman that the article seems in the end to lose contact with effect size, at times advising that power be treated "dichotomously" and rigidly irrespective of effect size.[2] "Important information can be found on the edges," as Rothman put it. But overall Rothman and we agree: it's a crushing piece. The oomph-laden content of their work is exemplary. Freiman and her colleagues note that the experiments and seventy-one oomphless, premature truncations were conducted by leading medical scientists. Such premature results were published in *Lancet*, the *British Medical Journal*, the *New England Journal of Medicine*, the *Journal of the American Medical Association*, and other elite journals. Effective treatments for cardiovascular and cancer and gastrointestinal patients were abandoned because they did not attain statistical significance at the 5 percent or better level.

SALMONELLA SIGNIFICANCE

On September 4, 1990, the South Carolina Department of Health and Environmental Control confirmed the existence of two people hospitalized with salmonella poisoning. The two had attended a hardware convention in Greenville, and both had eaten from a catered buffet lunch serving turkey and ham.[3] Of more than 398 conventioneers interviewed by state health department and Centers for Disease Control agents following the tipoff, over 34 percent reported gastroenteritis. Cultures confirmed that 9 of the initial interviewees had contracted the poisoning. The inquiry revealed an outbreak large enough to continue with a full-scale study of the event. In the study the statisticians and epidemiologists from the CDC and the state health department estimated that of 2,430 conventioneers 824 fell ill. This was, they said, "the largest food-borne outbreak ever reported in the state of South Carolina" (Luby, Jones, and Horan 1993, 31). It made the national news.

A kitchen employee admitted that some of the turkey—note: the turkey—was sitting the Saturday before the Sunday lunch for hours at room temperature "on the table." It was then shipped by unrefrigerated truck one hour north to another kitchen in North Carolina, cooked, and then reloaded on Sunday morning on another unrefrigerated truck. The truck broke down. Still another unrefrigerated truck arrived later and completed the delivery. Waitstaff working at Sunday's convention lunch complained about the malodorous turkey. Management ordered it reboiled in water—water that had gone tepid from boiling other turkey. To hide the smell the reboiled turkey was then cold rinsed before being laid on serving trays and set next to the ham on the buffet line. Note: ham.

It wasn't the first time the restaurant had sought cold-water rinsing as a solution to bacterial infection of food. The same restaurant had been penalized for similar behavior in several previous cycles of regulation.[4] The wonder is that no one died.

But the conventioneers ate ham, too, which was offered side by side with the turkey and was treated to the same lax preservation. *Ham eaters got sick, too.* The turkey consumption produced the greatest relative risk, the authors of the study found. "People who ate turkey were 4.6 times more likely to become ill than those did not eat turkey [95 percent confidence interval 2.0, 10.6]" (Luby, Jones, and Horan 1993, 31). You can see that the authors of the study were properly concerned with the magnitudes of effect. Sunday's cold-rinsed turkey eaters definitely stood out to them in the odds ratios—at 4.6-to-1.0 compared to those who did not eat the turkey. Size matters, the authors know, and also the confidence intervals, and they emphasized both. That is, they did for turkey.

But ham dinners, they argued, were not a culprit, because on the ham variable "$p < 0.17$," merely. Yet ham eaters suffered attacks in large numbers, too, and at an odds ratio high enough to matter medically. The authors ignored the finding. After all, the ham effect was statistically "insignificant." Shades of Vioxx. And on the same Fisherian grounds they ignored the results of their study of the *Saturday* lunch. Luby, Jones, and Horan admitted in the conclusion to their study that turkey was not the only mishandled product and Sunday was not the only bad day. But only the Sunday turkey got the headlines.

SIGNIFICANT DEPRESSION RELIEF

St. John's-wort is a perennial herb that blossoms annually on many-branched bushes into hundreds of yellow and star-shaped flowers. It has been used for millennia as a diuretic. And it has been used to treat nerve damage and to relieve bouts of minor and major depression. The efficacy of the "sunshine herb" is frequently under attack (perhaps, one suspects, because it seems to be a cheap substitute for drugs).[5] Two hundred patients with "major depression" were recently studied by Shelton et al. (2001). Each patient was randomly assigned St. John's-wort or a placebo (Rothman 2002, 124). The exact division was 102 to the placebo, 98 to St. John's-wort. (One wonders what the authors would think if a placebo taker killed herself while on it. But that is a different issue.) The authors believed that the less severely depressed patients at the time of their entry into the study would likely see the best results from St. John's-wort, where "best" means "highest relative risk" of going into remission. In the Shelton et al. study, the null hypothesis of "no effect" was defined to be a relative risk ratio of 1.0. To the extent that the relative risk ratio exceeded 1.0, then, St. John's-wort could be said to reveal its beneficial properties. Likewise, a ratio of 0.9 or 0.6 would be indicative of a negative effect. The results of the study, including exact p-values and the "relative risk" of remission, were published in the *Journal of the American Medical Association*. The key result, summarized by Rothman (2002), and is reprinted in table 16.1. The relative risk of "remission"— defined as a retreat of the symptoms of depression—was estimated to be in favor of St. John's-wort, relative risk = 2.0, but the p-value on the ratio was 0.14, statistically "insignificant," the authors said, by the Fisher and journal standard of 5 percent. Sheldon et al. concluded from the p-value that St. John's-wort is not clinically effective. Doesn't help, they said.

But, as the data in table 16.1 show, St. John's-wort is on average twice as helpful as the placebo. The data suggest that the flower does bring some sunshine, or at least diminishes the darker side of gray, and maybe by a lot. Rothman computed a p-value function—a continuous function of p-values

TABLE 16.1. St. John's-Wort on Trial:
Remissions among Patients with Less Severe
Depression (the Shelton et al. data)

	St. John's-Wort	Placebo
Remission	12	5
No remission	47	45
Total	59	50

Risk ratio = 2.0
90% CI 0.90–4.6
N = 200

Source: K. J. Rothman 2002, 125, table 6.4.

mapped against a range of effect sizes. The range of effect sizes was here again measured by the relative risk ratio and includes both beneficial and nonbeneficial effects. He shows that another hypothesis, a fantastically beneficial risk ratio, RR = 4.1, shares the same p-value, .14, as the null, RR = 1.0 (2002, 125). This is common in medicine and all the sciences. To think that p-values have a 1-to-1 correspondence with a unique risk ratio is to ignore the symmetry of the p-function.

Compare figure 16.2, which concerns an internationally publicized study purporting to show that aspirin does *not* reduce women's likelihood of having a major cardiovascular event. The aspirin study makes the same mistake as the St. John's-wort study. The scientists told women to stop taking aspirin even though, very clearly, the figure shows, aspirin was helping to save their lives.

Cancer Survival Is Significant

In 1995 some cancer epidemiologists made history. The authors of ten independent and randomized clinical trials involving thousands of patients in treatment and control groups had come to an agreement on an effect size. Consensus on a mere direction of effect—up or down, positive or negative—is rare enough in science. After four centuries of public assistance for the poor in the United States and Western Europe, for example, economists do not speak with one voice on the direction of effect on labor supply exerted by tax-financed income subsidies.[6] Medicine is no different. Disagreement on the direction of effect—let alone the size of effect—is more rule than exception.

So the Prostate Cancer Trialists' Collaborative Group was understandably eager to publicize the agreement on a *real* standard of significance. Each of the ten studies showed that a certain drug, "flutamide"— for the treatment of prostate cancer—can increase the likelihood of patient survival by an average of 12 percent (the 95 percent confidence interval in the pooled data put an upper bound on flutamide-enhanced survival at about 20 percent [Rothman, Johnson, and Sugano 1999]). Failure odds of 5 out of 100 is not the best news to deliver to a prostate patient. But if castration followed by death is the next best alternative, a non-invasive 12 to 20 percent increase in survival sounds good.

But in 1998 the results of still another, eleventh trial were published in the *New England Journal of Medicine* (Eisenberger et al. 1998, 1036–42). The authors of the new study found a similar size effect. But when the

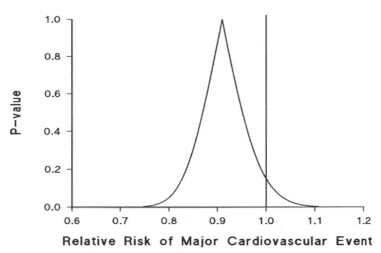

Fig. 16.2. Likewise, Rothman shows, aspirin saves women's lives even when the aspirin is "statistically" insignificant. The vertical line is located at the null hypothesis, Relative Risk = 1.0, "no effect." Notice that the empirical *p*-value function lies almost entirely on the side of "aspirin works" (the risk of heart attack is less for aspirin takers than for non–aspirin takers). Authors Paul Ridker et al. (2005) said they could not reject the null hypothesis of RR = 1.0, however, since the *p*-value for the hypothesis is about .17. "It was a nonsignificant finding with respect to the primary end point," they wrote. But the same *p*-value obtains for RR = .85—an aspirin-induced reduction in heart attack risk of 15 percent. (Source: Kenneth Rothman, personal communication, Boston University.)

two-sided *p*-value for their odds ratio came in at .14 they dismissed the efficacious drug, concluding that there was "no clinically meaningful improvement" (1036, 1039). Kenneth Rothman, Eric Johnson, and David Sugano examined the individual and pooled results of the eleven separate studies, including the new study conducted by Eisenberger et al.

> One might suspect that [Eisenberger et al.'s] findings were at odds with the results from the previous ten trials, but that is not so. From 697 patients randomised to flutamide and 685 randomised to placebo, Eisenberger and colleagues found an OR [odds ratio] of 0·87 (95% CI 0·70–1·10), a value nearly identical to that from the ten previous studies. Eisenberger's interpretation that flutamide is ineffective was based on absence of statistical significance. (1999, 1184)

Rothman and his coauthors depict the flutamide effect graphically in a manner consistent with a Gosset-Jeffreys-Deming approach. That is, they pool the data of the separate studies and plot the flutamide effect (measured by an odds ratio or the negative of the survival probability in a hazard function) against a *p*-value function. Using the graphical approach Rothman and his coauthors are able to show pictorially how the *p*-values vary with increasingly positive and increasingly negative large effects of flutamide on patient survival. And what they show is substantively significant.

> Eisenberger's new data only reinforce the findings from the earlier studies that flutamide provides a small clinical benefit. Adding the latest data makes the *p* value function narrower, which is to say that the overall estimate is now more precise, and points even more clearly to a benefit of about 12% in the odds of surviving for patients receiving flutamide.

Rothman, Johnson, and Sugano conclude that "the real lesson" from the latest study is "that one should eschew statistical significance testing and focus on the quantitative measurement of effects."

That sounds right. Statistical significance is hurting people, indeed killing them. It is leaving their illnesses and a defective notion of significance "unexplained."

17

Edgeworth's Significance

Sir John Lubbock used to nick the wings of his wasps and bees in order to observe the time that each took in performing certain operations: namely (1) taking in a load of honey, (2) carrying it to the nest or hive and returning. The trouble of marking individuals may sometimes be avoided where it is possible to observe . . . those who are beginning and those who are ending an operation. Of this character is the problem to find the average time of a wasp's absence from the nest, by merely counting the numbers going into and out of the nest from time to time.

EDGEWORTH 1907, 47

How could such a strange scientific turn have taken place? How could it be that so many sciences became sizeless and ($p < .05$) hazardous to health and wealth?

Francis Ysidro Edgeworth (1845–1926), the inventor of the very word *significance* in its statistical meaning, is not to blame. When in 1885 Edgeworth coined the term he was studying a sample of bees and wasps he had personally observed and collected. Edgeworth, an Oxford man, came from an astonishing family of Anglo-Irish gentry. He became effortlessly a fellow of All Souls and lived quietly as a prolific writer on this and that, earning distinction in philosophy, economic theory, and mathematical statistics. He read Latin and Greek with ease and would sprinkle his writing on the hedonic calculus and the asymmetric probability curve with classical allusions. He could dash off examples of "significant differences" in fruit farms, planets, death, or money and did so in his 1885 article "On Methods of Statistics." In print Edgeworth could argue with the gods.

EDGEWORTH COUNTED WELL

In person he was shy, living like a monk. He did not keep much of anything—not even books, preferring instead to use libraries. Eyes at half-mast, he was withdrawn from the human swirl, a sweet man lost in thought. He sat with humans as little as possible. For hours and for days, however, in England and in Ireland, he sat with wasps and bees.[1]

He wanted to know for how long on average the wasps and bees would abandon the businesses of sex and work for other activities. It was an economist's sort of question. How much leisure time would they take for dreaming or for thoughtless bullying, for simply random wandering? He wondered which variables—time of day, for example, or outside temperature—would make the insects return home quicker. Edgeworth's calculations showed that not all insect behavior is identical. Some behaviors are random, some not—a matter of inside temperature, of character, for example. There's the average *Apis* and *Vespula,* and then there's the exceptional. One is reminded of Alfred Kinsey decades later moving from the variability of gall wasps to the variability of sexual behavior in the human. Kinsey and Edgeworth followed Virgil: *parvis componere magna,* "to compare great things with small."

In 1885 significance testing was the sweetest instrument Edgeworth could think of with which to "discriminate" (1885, 209) quantitatively the random events from the more permanent character traits, to distinguish, as Edgeworth himself put it, "between material and accidental oscillations" (209). Statistical significance was not worked out fully by Edgeworth (Stigler 1986, 322). Like Quetelet and the astronomer G. B. Airy before him, Edgeworth found "significance" by dividing observed differences—usually of means—by the "modulus"—in today's terms, the square root of two times the standard deviation. He made little distinction between probable errors in small and large samples, and he provided no theory for setting the odds on null or alternative hypotheses. He asserted mildly that a result "two or three times the modulus" might not be "accidental." He did not think at all about alternative hypotheses.

Such oversights were in 1885 of course not scandalous. Essentially no one thought about alternatives to the null hypothesis until Gosset did in the crucial letter to Egon Pearson in 1926. In fact the "null" itself wasn't thought of either, except tacitly—Fisher routinized that, too, in 1935, as a part of his battle against prior probability. And the mathematical formalization of "power" and "loss," inspired by Gosset, was in 1885 still forty years away.

What mattered to Edgeworth in all his studies, however, was the "material" oscillation, as he put it, the "*substantial* difference" (1885, 206; italics supplied)—the oomph, as we would say, measured by the social or economic implications of a bee's time away from home or by price inflation induced by a larger circulation of commercial bills or by death caused by tuberculosis. To Edgeworth size did matter. He always wanted to know How Much. To Edgeworth, who was a broad-gauged scientist with a feel for cost, an observed difference in price or honey or life was important if it was "enough [of a difference] to require the continuation of the inquiry"—"if the subject," he meant, economically or experimentally or ethically speaking, "repaid the trouble" (208). Note his economistic way of putting it.

A statistically significant difference may be immaterial, Edgeworth observed, worthless for policy or science, not enough to warrant further discussion. And a statistically *insignificant* difference may on economic or clinical or ethical grounds—despite the buzzing noise around its estimate—"repay the trouble" of continuing the inquiry. In other words, the coiner of the term *statistical significance* knew that oomph is what mainly matters.

Consider an extended example from his seminal article: "Ex[ample]. 3 . . . taken from a species of industry which has not received much attention from statisticians . . . the image of trade which is presented by wasps entering and issuing from their nest" (208). "The exports and the imports fluctuate with remarkable regularity," Edgeworth immediately observed in his estimate of total traffic, "mean total traffic (exports + imports)" (209), without the aid of a significance test (see table 17.1). In the article he presented "observations" on wasps "made by me at Edgeworthstown, Ireland, September 1884. They were made on different days at different hours. They all relate to the same nest." Since Edgeworth had collected his own data, he knew his observations intimately; for example, he controlled exactly for nest- and time-of-day heterogeneity, reducing error in observations that cannot be matched with a mere test of statistical significance on a data set downloaded from the Internet, no matter how mathematically advanced the "correction." He of course exhibited, unlike many articles in the *American Economic Review* and the *New England Journal of Medicine*, the actual levels of the variables with which he was concerned.

Table 17.1 reproduces Edgeworth's table. The first column denotes "the number of minutes" (1885, 209) he observed wasps leaving and entering the nest. The second column, titled "Mean," is "total traffic (exports + imports) per minute." And the third is "the modulus . . . by the Method of Mean Square of Error"—"the square root of the mean fluctuation" (209), that is, as we said, the square root of two times today's definition of the estimated standard deviation.

Let Edgeworth summarize his own results. Observe the size-matters/how-much thinking as against Fisher's existence/whether thinking.

> It is clear that for that nest at that place and season, and for a traffic of from 25 to 30 per minute, the modulus 8.5 (the square root of the mean fluctuation) might safely be employed to discriminate between material and accidental oscillations. For example, on 4th September at 8 a.m. and at noon, the mean total traffic was respectively 42, 40; the corresponding observations numbered 5, 13. If in an insect republic there existed theorizers about trade as well as an industrial class, I could imagine some Protectionist drone expressing his views about 12 o'clock that 4th day of September, and pointing triumphantly to the decline in trade of 2½ per cent. as indicated by the latest returns. Nor would it have been easy to refute him, except by showing that whereas the observed difference between the compared Means is only 2, the modulus of comparison is $\sqrt{70/5 + 70/13}$, or 4 at the least; and that therefore the difference is insignificant.
>
> The gross fact that a decline of temperature is frequently accompanied with a decline of trade is obvious to ordinary observation. . . . It is an interesting question, whether in this miniature trade there is an excess of imports [buttressing the prejudice of the Protectionist drone]. The

TABLE 17.1.   Edgeworth's Observations of Trade among Wasps, September 1884

"Number of Observations" (number of minutes wasps observed)	"Mean" (exports + imports)	"Modulus" (square root of mean fluctuation)
22	26	7
21	24	10
22	18	8
17	33	8.5
20	26	10
16	25	8
13	40	7
21	42	9

Source: Edgeworth 1885, 209.

modulus having been ascertained, both for imports and exports—the latter appears to be the greater—it is found that there is no such excess. The imports are paid by the exports. (1885, 209–10)

Edgeworth slides into ambiguous usage of the very word *significance* here (first paragraph, last line), as he does in fact throughout the article, meaning by the word sometimes the narrowed statistical sense only (contrast Edgeworth 1907, a richly empirical article that avoids the ambiguity). That's too bad. But ambiguous usage is only one of a score of misfires on the point in later usages of the sizeless sciences and is hardly the most damaging among them. Edgeworth's concern with the substance comes out clearly. He emphasizes the levels and movements of his key variable. He compares his results to another sample, his August 1885 study of four other nests (1885, 209ff.). He evaluates the magnitudes with the aid of economic theory, not with the aid of a sizeless switch, significant/insignificant. And he answers reasonable doubt by making a further calculation, an accounting calculation on the wasp-nest balance of payments. "It is found that there is no such excess [of imports]."

Edgeworth says elsewhere in the article, in a different example, that according to the means "no great difference" separates the amounts of money bills in circulation in different quarters, 1830–53, yet "still slight indication of a real law—enough to require the continuation of the inquiry, if the subject repaid the trouble" (1885, 208). We can take Edgeworth to mean "repaid" in an economic or other substantively significant sense (contrast Hoover and Siegler 2008, 4, fn. 2).

Statistical significance can—assuming all the other errors that can and do arise in science have been corrected for—indicate the likelihood of the presence of an effect. At an arbitrary level of likelihood it can be what Hoover and Siegler (2008) and we would call "a signal." But, says Edgeworth, and with him McCloskey and Ziliak, so what? Remember Thoreau about the telegraph. Whether a lot of little private banks or one great big central bank should do anything about the small difference depends on whether the subject, economically speaking, repays the trouble. Hoover and Siegler want to assign the responsibility to a man they call "practical." Shades of Fisher: the scientist is replaced by a mechanical puppet who acknowledges a signal at $p = .05$, and the puppet—not the scientist who knows why it might matter—is called "practical."

Edgeworth consistently defended a size-matters/how-much approach to quantitative judgment. The 1885 article ends with a warning about sign econometrics: "(1) where in each comparison we look only to the fact of excess or defect [of wasp trade], and (2) where we take account also of the *extent* of difference according to the principles. . . . For example, in comparing [say, the heights of] several groups of men under 30, we may . . . look only at the *sign* of each difference, or we may also take account of the *quantity*. The first process [looking at the sign] requires no illustration. It is only too familiar. For it is to be feared that many statisticians have not got beyond this operation" (213; italics in original).

Somewhat impertinently we applied our nineteen-item questionnaire to Edgeworth 1885: it scores sixteen of nineteen Yes. Given that power and power functions were not operationalized until the late 1920s, Edgeworth's article of 1885 would actually score sixteen of seventeen possible Yes, putting it in the ninety-ninth percentile of the *American Economic Review,* 1980 to the present. We are not surprised. Economic significance was Edgeworth's concern "from," as he puts it characteristically, "morn to noon, from noon to dewy eve" (209; with reference to *Paradise Lost,* 1:742–73; cf. the Iliad, 1:591ff.).

Decades later Edgeworth was still troubling with wasps and bees (Edgeworth 1907). He liked them. Life was sweet on Hampstead Heath, peace dropping slow. One evening in September 1906 Edgeworth's sample of dead wasps, taken with "spray" and "blows" to a nest in Hampstead, had grown large enough. Day bags were heavy, and back home sat several scientific articles waiting to be proofread. Time to seal up the hole in the nest. Edgeworth would return in the morning to open up that same hole. He wanted to see how the remaining wasps would recover from the loss of workers and of the freedom to migrate that he had caused (366). An economist at heart, he wanted to know how elastic wasp behavior was with respect to the exogenous losses. Would wasp behavior change? By how much?

18

"Take 3σ as Definitely Significant":
Pearson's Rule

*Take words, graphs, maps and symbols. . . . They are never objects
of our attention in themselves, but pointers towards the things they
mean. If you shift your attention from the meaning of a symbol to the
symbol as an object viewed in itself, you destroy its meaning. . . . The
skillful use of a tennis racket can be paralyzed by watching our racket
instead of attending to the ball and the court in front of us.*

MICHAEL POLANYI 1959, 30–31

The argument from odds is old. Stephen Stigler notes that one might read
a significance test into the clinical trial in the Book of Daniel (1:12–16),
though "[i]t may be a stretch" as "the significance level and even the test
statistic were left vague there."[1] We have mentioned Cicero on the matter. In 1711, John Arbuthnot presented a paper to the Royal Society of
London proposing to show statistically that "Divine Providence"—not
mere chance—governs "the constant regularity observed in the births of
both sexes." A few years later Daniel Bernoulli conducted tests of significance on the hypothesis of the randomness of planetary orbits. And in
1773, we have mentioned, too, Laplace tested the hypothesis that comets
come from outside the solar system.[2]

The very notion of a single, "crucial" test was itself, we have noted,
part of a neopositivist notion of scientific method, that is, new to science.
A nineteenth-century gentleman would look at all the evidence, a notion
as we have noted embodied in John Stuart Mill's *System of Logic* of 1843.
Edgeworth himself was a confirmationist and perspectivalist. Falsification
and a single test was, he thought, for amateur empiricists. He dreamed of

one day employing a psychic utility machine—a "hedonometer"—to get copious, direct readings on the human experience of pleasure and pain (1881). We appear to be close to such instruments in modern scans of the brain. But short of direct measurement, inferential statistics appeared to be necessary. And inferential statistics meant odds ratios.

Probability Was All Karl Could Know

Karl Pearson was born Carl in London, in 1857, the second of three children of a barrister. The Pearson family lived north of London until 1866, when they found a house on Mecklenburgh Square in Bloomsbury, the neighborhood where Carl would work and live for much of the rest of his life. (There seems to have been no conversation between the statisticians close to Carl and the later Bloomsbury group, which met literally around the corner from Gower Street, though both groups were at their intellectual height during the 1920s and Keynes could have functioned in both.) Carl changed his name to Karl in 1880 in honor of his years of study in Germany. A year or two of study in Heidelberg or Berlin was the custom for Englishmen and Americans at a time when Germany's universities exceeded by far the standard of graduate work at home. (Think, for example, of DuBois.) Karl was a Germanophile all his life. The Englishman was offered a chair of Germanic literature at Cambridge and named his children Sigrid, Helga, and Egon.

In his youth Pearson ranged through mathematics and history and literature. He had the breadth of an Albert North Whitehead, of whom it was said "Whitehead knows both"—that is, both sciences and arts. Pearson, too, knew both. He graduated from Kings College, Cambridge, in 1879, Third Wrangler in the Mathematical Tripos. He took up with fervor the subjects of physics, feminism, and socialism.[3]

The historian of statistics Theodore Porter has noted the tension with which Pearson struggled in the years before the publication of *The Grammar of Science* (1892) between science and passion, ethics and observation. "He came before the world as a distinctive scientific persona," writes Porter, "but he worried without end about the implications of science for the person—one might almost say the soul" (2004, 9). Pearson chose the scientific side, he thought, by turning, under Galton's influence, to a mechanical philosophy of inference. As Porter writes,

> As a young man, [Pearson] recorded several ecstatic encounters with nature, which he regarded as among the most powerful moments of his

life. These were notably unscientific experiences [he believed], and his subsequent identification of knowledge with numbers and cool precision was a real act of renunciation for him. In *Grammar,* he made rationality stand for distance from objects of desire, knowledge being only of sensations or appearances. (10)

But the Cambridge-trained mathematician nonetheless wrote discursive cultural histories of Germany and Europe. He studied folklore and worried and rejoiced over the likely meanings of poems and parables. He immersed himself in Ibsen, Strindberg, and Shaw and counted Shaw as a friend. He published a passion play. He published the story of his storm and stress in an autobiographical play, *The New Werther,* modeled on Goethe's preromantic romance, *The Sorrows of Young Werther.*[4] Around 1890 Pearson was fervently searching for a release from this torturous dialectic. He longed, in Ian Hacking's phrase, "to tame chance" (Hacking 1990). Pearson seems to have been a romantic who wanted to be classical.

And then probability theory and its bourgeois cousin, applied statistics, descended, with encouragement from Galton, like a new "evangel." Knowledge of "sensations" and "appearances" would come not from poetry or pushpin but from probabilistic thinking and statistical measurement and inference—that is, knowledge would come—pace Bruno de Finetti—from a distance. The metaphysical "thing in itself," the inheritance of Kant, is dead, Pearson believed. Causation is a limit point in a world of correlation, itself a probability.[5]

"Statistics was the answer to [Pearson's] quest for a life mission," Porter writes, "a field defined by methodology and mathematics that licensed him to make incursions into every man's specialty" (2004, 9). As Porter suggests, it was satisfactory for a compendious intellect like Pearson's. The work to be done in statistics especially suited his temperament. Pearson was from 1884 to 1911 Professor and Chair of Applied Mathematics at University College London on Gower Street. In 1911, when Francis Galton died, Pearson was appointed Galton Professor of National Eugenics, a gift of his friend Galton.[6] It was the same chair, with the Eugenics Laboratory attached, that Fisher was eventually to receive. In the half century before 1939 racist applications of statistics were British territory and Galton-Pearson-Fisher was the high road through it.

When Pearson embarked on bringing mathematical statistics into the mainstream of science he was already famous as the thirty-five-year-old author of a neopositivist blockbuster, *The Grammar of Science. The Grammar* was among other things a response to another foundational grammar, Cardinal Newman's *An Essay in Aid of a Grammar of Assent* (1870). Newman's book had considered what Egon Pearson would later call Gosset's "second kind of error"—that is, gullibility, the pejorative of faith. Newman called it the "illative sense"—the positive and rational need to believe, to act, to give assent, even in the face of uncertainty. Karl Pearson's project ignored the second kind of error—Newman's first kind of wisdom—and elevated doubt.

But mainly *The Grammar* was Pearson's attempt to unify the epistemology of the sciences while resituating the scientist in the wider society. To Pearson probability was the only thing we could reasonably know. We are not to focus on the thing itself, or its relation to other things, which are unknowable directly, only on their probabilities. His replacement of statements about things with statements about the probabilities of things intrigued a young Wittgenstein. When Einstein the patent clerk became involved in a new reading group in Bern its first book was *The Grammar of Science*. About the same time some students and faculty at the University of California-Berkeley started up a Pearson Club. Lenin was impressed by the book's "philosophical consistency." A young Jerzy Neyman, studying mathematics with Bernstein at the University of Kharkov in Ukraine, was deeply struck by Pearson's book:

> We were a group of young men who had lost our belief in orthodox religion, not from any sort of reasoning, but because of the stupidity of our priests. . . . *The Grammar of Science,* which Bernstein recommended, was striking, because (1) it attacked in an uncompromising manner all sorts of authorities and (2) it was the first attempt we had to construct a "Weltanschauung," not on any kind of dogmatic basis, but on reason. (Neyman n.d., in Pearson 1936, 213–14)

Pearson's *Grammar,* Neyman said, had sustained him and his friends through Red October (214). Henry Adams, for his part, declared that "the rise or fall of half-a-dozen empires interested the student of history less than the rise of *The Grammar of Science.* . . . Pearson had destroyed [with one book] the order of nature, leaving a chaos of chance, and had reduced 'truth' to 'a medium of exchange'."[7]

The objects of science themselves—profits, planets, pain, prescriptions—existed in the realm of passion or ethics, at any rate so far as humans were involved in viewing them. But passion and ethics are no guide to science, Pearson believed, and certainly did not achieve the objectivity for which he longed. Pearson did not accord legitimacy to introspection—

contrast here Descartes, Kant, Newman, and the Bayesian statisticians. A scientist could only objectively reason about the probability of a numerical relationship between, say, pain and some prescription pill for killing it—something Pearson believed to sit outside the viewer and the viewed.

His view was at once short and far. Inference was to him not a property of purposive human action or of belief but a piece of information about the frequency of occurrence of things outside the viewer, what Fisher would later call "Natural Knowledge" (1956, 103). Since statistical significance as understood by Pearson in the 1890s was a measure of the first kind of error, so to speak—that is, attentiveness to skepticism rather than faith—Pearson had found his major scientific instrument: statistical significance. To speak in regression model terms, after *The Grammar* and his probabilistic leap of faith, Pearson's gaze shifted from the quantitative relationship between y and β, that is, from the magnitude of the effects about which one is to assent, toward α and ε, that is, the probability of error about which one is to doubt, then falsify. His gaze was fixed in this way for his three decades of editorship of *Biometrika*. The lifelong antipathy of his intellectual heir, Ronald Fisher, to cost functions, Bayes's rule, the second kind of error, and the "acceptance" of hypotheses was justified in *The Grammar*. It justified a substitution of doubt for belief and a scientific stance of disinterestedness rather than civic participation.

Decades later, in *The Counter-revolution of Science*, Friedrich Hayek attacked such a distant, mechanical, and scientistic approach.

> Whether it is the conception of an observer from a distant planet, which has always been a favorite with positivists from Condorcet to Mach [and from Pearson and Fisher to recent neo-positivists], or whether it is the survey of long stretches of time through which it is hoped that constant configurations or regularities will reveal themselves, it is always the same endeavor to get away from our inside knowledge of human [or other] affairs and to gain a view of the kind which, it is supposed, would be commanded by somebody who was not himself a man but stood to men in the same relation as that in which we stand to the external world. . . . In most instances this belief that the total view [such as the view allegedly afforded by the Pearson-Fisher view of statistical significance] will enable us to distinguish wholes by objective criteria, however, proves to be just an illusion. (1952, 59)

Illusion or not, Pearson's worldview found a powerful institutional outlet. In 1901 Pearson launched with three other celebrities of science— his mentor Galton himself, the Oxford biologist William Weldon, and the

American leader of eugenics, C. P. Davenport—the journal *Biometrika*. A few years later, with financial support from Galton and the London Worshipful Company of Drapers, the three founded the Biometric Laboratory, the journal's think tank. Pearson was annoyed by Edgeworth's antisocial resistance to joining them.

Statistics in the English-speaking world did not then exist as a distinct science—the French and Germans were in some ways ahead. Pearson started a Department of Applied Statistics at University College, the first of its kind. At first a nondepartmental, administrative unit, providing an umbrella for all of Pearson's projects, the unit emerged as a distinct Department of Applied Statistics just after World War I. In the United States, we have noted, the first department of statistics was formed in 1935 by a Fisher disciple, George Snedecor, at Iowa State University. But *department* doesn't convey enough in either case. It was largely *Biometrika,* which Pearson edited from 1901 to 1936, and his two think tanks of Gower Street, the Biometric Lab and the National Eugenics Lab, that created today's separate science of statistics. The Iowa State and Stanford departments (where Hotelling held forth) were important first imitators.

PEARSON HAD OTHER FISH TO FRY

Statistics was not by any means the primary science on the Gower Street agenda. Biometry, but especially eugenics, was. In the inaugural issue of *Biometrika* the editors used Darwin as a springboard to justify "biometry" and "eugenics," their "other" new and separate sciences.[8] "The first condition necessary, in order that any process of Natural Selection may begin among a race, or species," Pearson and his fellow editors wrote, "is the existence of differences among its members"—as Hayek philosophically put it, the distinguishing of wholes by objective criteria. Molecular biologists would later demolish the logic of Pearson's "first condition." But Pearson's emphasis on the phenotypic extremes of his subjects, from wrinkled peas to Jewish eyes, was, historically and progressively speaking, the post-Darwinian leader of biometrics believed, both a natural historian's proof of Darwin's natural selection and a statistical attack against a still-pervasive essentialism.

> The unit, for which such an enquiry must deal, is not an individual but
> a race, or a statistically representative sample of a race; and the result
> must take the form of a numerical statement, showing the relative fre-

quency with which the various kinds of individuals composing the race occur. (Pearson et al. 1901, 1–2)

Pearson's papers and the archives of the Biometric and Galton labs survive. One finds in them the ephemera of a scientific racism common to the age, and to which Galton, Pearson, and Fisher were leading contributors: photographic plates of black albino penises, shadowy photos of young boys jumping distances, long discursive and statistical treatises on the quality of Jewish eyes and the sizes and shapes of native Indian skulls, Spearman's first measures of IQ.[9] "Race," Pearson wrote in his new journal, was "the unit, for which such an enquiry [as biometry and eugenics] must deal . . . and the result must take the form of a numerical statement, showing the relative frequency with which the various kinds of individuals composing the race occur" (Pearson et al. 1901, 1–2). Value judgments—arguments about the arguments—and Gosset's personal probability, were to be kept out of the neighborhood of their new sciences. Pearson would write in the 1920s against Jewish migration to Britain, and Fisher would write in the 1930s against material relief for poor people and literally in favor of relief for the rich on eugenic grounds. Such stuff was in the air and, as we say, even among the leading white progressives, from G. B. Shaw to the economist Richard T. Ely and Karl Pearson himself.

So *Biometrika* attracted top statistical minds. The first decades of the journal read like a who's who. Statistical giants, in fact, walked Gower Street: first and long before was a great collector of data, Charles Darwin, who had in fact lived in a house at 17 Upper Gower Street, then Galton, Pearson, and Pearson's mighty recruits: G. Udny Yule, Alice Lee, Florence N. David, Charles Spearman, Major Greenwood (a pioneer in medical statistics), and, of course, Gosset, Egon Pearson, Neyman, Shewhart, and R. A. Fisher. (Statistical giants, by the way, still can be seen on Gower Street: Dennis Lindley took over the Statistics Department in the 1970s and is now Emeritus Professor at University College.)

THE SIZELESS STARE ORIGINATES WITH PEARSON'S RULE OF THREE

Pearson longed for objectivity. But by eliminating judgment in science he undermined his other fervent wish, to promote the development of the engaged scientist-citizen, most of all K. Pearson. He sought probability rules. "Now I suggest," he told first-term students, "that the fundamental problem of statistics is this: I observe an event in N trials to succeed P times and fail Q, what is the probability that in M further trials it shall succeed R times and fail S, the circumstances of the trials remaining the same?"[10] In his 1905 lectures on "Fundamental Conceptions," the "success" and "failure" were said to be matters of sampling precision alone, and came furnished with a rough and ready rule, to quote again Pearson's lecture notes:

> *Probable errors of constants. Immense importance of.*
>
> How much deviation from expected value to be consider[ed] so improbable as to discredit result? This is a matter of experience & practical working.
>
> Thus we may take:
>
> | | |
> |---|---|
> | < 2 p.e not definitely significant | 1.35σ |
> | > 2 < 3 p.e possibly significant | 2.02σ |
> | > 3 p[.]e < 4 p.e probably significant | |
> | > 3 p.e almost certain significance | 2.70σ |
>
> (Pearson ca. 1905, 1; italics in original)

The credibility of a result, he told his students, was a matter of probable error—that is, the level of statistical significance. Pearson appears to correct himself: "Of course want of significance does not mean that there is not real differentiation. It may only mean that the data are too sparse to give any result" (1). But one cannot easily take Pearson here to mean that size matters or that alternative hypotheses matter or that the loss function matters. His point is merely that the sample is too small to attain the blessed 3 p.e., a fact proved from Gosset. "Real" to Pearson did not mean "having oomph, being scientifically important, tempting the will." It meant 3 p.e.

Pearson further explained his philosophy with examples taken from studies of employment, eye color, death rates, and "the cephalic index in man" showing that errors of other kinds—such as, for example, what we now call "omitted variable bias"—should be controlled. "If a sample be taken of 1000 . . . this year from a particular species, it does not follow that it will be identical with 1000 next year within the limits of random sampling. The supply of food may have been different, the environment more favorable and this enacted on the character we are considering. . . . Hence one of the fundamental points will be to enquire when two random samples differ by more than the differences due to their being random samples—we shall then be able to assert definite change" (1905, 1–2). "Definite change" did not mean "mattering for policy, large according to some loss function." It meant: 3 p.e.

In Pearson's way of doing statistics, once the model and experiment are properly designed (whatever precisely he meant by this), is to let the statistical significance applied to mere sampling error decide the rest. Pearson says, "[T]he "*formula* [for the probable error of a difference is] *most important* for example in testing local deviations from [the] general death rate, local variations from general physical or psychical [or] anthropometric characters, etc." (6 [1905, p. 1 of "Summary" section]; italics supplied). The magnitudes and the meanings of death and unemployment rates, in Pearson's way of looking at it, are not important—not, anyway, to the scientist in search of "objective" knowledge, a Karl quite different from the passionate student of drama and German literature.

Pearson had himself been misled by his teacher. As Galton put it in his "Risk of Misclassification, Part I" (1899), "h is the measure of precision; its reciprocal $1/h$ being the measure or modulus of fallibility, or, more briefly, the modulus. It is with this modulus that we shall chiefly be concerned. It will soon be seen that the only datum required for solving the first problem [of misclassifying an observation] or any other of its class, is the ratio between (1) the modulus of the variability of the objects that are to be classified, & that which will be called a and (2) the modulus of the fallibility of the examiner, which will be called b."[11] To Galton, scientific inference would be evaluated under no specific or general currency—just probable error. It is a plain statement of precision only. Failure and success were to be measured exclusively on a scale of sampling probability. Pearson, too, in his quest for scientific rigor, settled, like his friend and mentor Galton, for a philosophical Whether.

In 1906–7 William Sealy Gosset, thirty years old but already six years an "experimental brewer" at Guinness, attended Pearson's lectures on probability and statistics. Invited by Pearson, Gosset came to Gower Street for a year as a postgraduate student in the Biometric Laboratory on sabbatical. Sometime in 1906 Pearson himself—who was perhaps irritated by Gosset's talk of "pecuniary value"—would concoct a rule. Under the blackboard heading *Normal Curve,* K. P. told the very class that Gosset was sitting in what was to be considered "significant" and what not.

Prob. error: . . .

Looking in the table if α = .5 . . . 2 to 3 times the prob[able] error is not definitely significant but approaches significance as we get on to 3.

   4 times to give odds of 250:1

   3 -------------------- 50:1

   2 -------------------- 9:1

   1 -------------------- 3:1

Take 3σ as definitely significant

(*K. P.'s Lectures I,* 1906, unnumbered page 13,
Pearson Papers, Gosset file, UCL; final italics supplied)

And so the authors in *Biometrika*—save Gosset and Egon Pearson and Neyman and a few others—did as they were taught. Follow Karl Pearson and the Rule of Three. In 1901, 1.5σ was "definitely significant." By 1906 and for much of the ever after, authors published in *Biometrika* took "3 σ as definitely significant"—regardless of oomph, regardless of the meaning of the magnitudes uncovered in Pearson's other sciences.

19

Who Sits on the Egg of *Cuculus Canorus?*
Not Karl Pearson

An early case, applied to the eggs of the cuckoo bird, illustrates literally the feel of substantive as against statistical significance. "The Egg of *Cuculus Canorus*" (1901) by Oswald Latter was one of the first articles published in *Biometrika*. Edited closely by Pearson—Porter notes that Pearson's style of editing stretched the definition of "independent authorship"—Latter's article was a thoroughly Pearsonian product.

Cuckoos survive by stealing the domestic labor of others. They use other birds to sit on their eggs, sneaking their cuckoo eggs into the nests of the involuntary foster parents. "An explanation is needed," wrote Latter in an amusing locution, "of the success which attends this imposition." Professor A. Newton, in his *Dictionary of Birds,* had offered an explanation.

> Without attributing any wonderful sagacity to her, it does not seem unlikely that the cuckoo which had once successfully foisted her egg on a reed-wren . . . should again seek for another reed-wren's . . . nest . . . and that she should continue her practice from one season to another. . . . Such a habit could hardly fail to become hereditary, so that the daughter of a cuckoo which always put her egg into a reed-wren's. . . nest, would do as did her mother. . . . It can hardly be questioned that the eggs of the daughter would more or less resemble those of her mother. (Newton, quoted in Latter 1901, 165; spelling modernized)

The mother cuckoo's problem resembles the so-called principal-agent problem of industrial organization. Or as Dawkins see it, an "arms race" (Dawkins 1999, 55). How does she succeed?

She would succeed if her eggs *relevantly matched* those of the victim. Egg characteristics (size, color) and propensity to match *this* variety of cuckoo with *that* species of victim (reed-wren or whatever) would be selected for. In other words, Newton and Latter were saying, the foster mother can be duped by size *or* color. The word *duped* is metaphorical. By 1901 the Mendel-Darwin-and-biometrics debate had neared full pitch, and Pearson, Weldon, and Yule—the biometricians—were central participants in it. But in 1901 no faction or scientist believed "intent" was involved in the cuckoo behavior, merely unit-character inheritance or natural selection or both (Mayr 1982, chaps. 16, 17; cf. Dawkins 1999, chap. 4).

The trick for success—indicated by a hatch in the foster mothers' clutch and the carrying on of that set of genes—is to avoid imposing too obviously on the foster mother, the sitter. *But a "too obvious" imposition is something decided by the foster mother, not by an arbitrary limit of standard errors in egg dimensions observed by humans.* This is the point that Galton, Pearson, and Fisher entirely missed, right at the origin of the quantitative revolution.

An oversized egg would "inconvenience the sitter;" as Latter elegantly put it, and an oddly colored egg would alarm her. Therefore "my [statistical] enquiry," Latter explained,

> has thus resolved itself chiefly into an attempt to ascertain (1) if the eggs of cuckoos deposited in the nests of any one species *stand out as a set apart* from Cuckoo's eggs deposited elsewhere; (2) if the same eggs *depart from* the rest in such a direction to approximate in size to the eggs of that particular species of foster-parent. (1901, 166; italics supplied)

He continues—and here is the crux.

> The method employed is to find the mean (M) length or breadth, as the case may be, thence to compute the standard deviation . . . and then to find $100\sigma/M$, the coefficient of variation. *To test whether any deviation is [statistically] significant, M_r is taken as the mean of the whole race of cuckoos and M_s the mean of cuckoos' eggs found in the nest of any one species of foster-parent: the standard deviation σ_s of such eggs is also as-certained. . . . If the value of $M_r - M_s$ be not at least 1.5 to 3 times as great as the value of the other expression the difference of $M_r - M_s$ is not definitely significant.* (Latter 1901, 166; italics supplied)

Here is what Latter is arguing: he observes that cuckoo eggs are typically larger than other bird eggs in size—in length and breadth. This is prior knowledge and suffices to set the test, in Fisher's terminology, as one-sided. Cuckoo eggs in the foster family's clutch tend to be smaller

than cuckoo eggs on the whole. That is, cuckoos would presumably impose on the bigger of the smaller-egg species. Latter examined 223 of 243
cuckoo eggs that successfully stayed in the nests of foster parents of fully
forty-two different species of victim (1901, 166). If the ratio of the difference of egg size to the standard deviation—that is, if the sample mean
egg size of the cuckoo population minus the sample mean size of the
cuckoo eggs in the foster clutch, expressed as a ratio of the standard
deviation of the cuckoo eggs in the foster clutch—*meets or exceeds* "1.5,"
then Latter (and Pearson) interpret the signal-to-noise ratio as "definitely
significant" and "conclusive" (1901, 166) in support of Newman's theory.
That is Pearson's Rule of Three, one sided (1.5 times 2 = 3.0).

Latter and Pearson employed only a standard of precision *without asking whether the projection caused by the addition of the large foreign egg
would "inconvenience the sitter."* Latter and Pearson did not draw a line
of oomph for cuckoo eggs. What matters to mother reed-wren (or mother
meadow pipit or mother reed warbler) is, to repeat, the *size* of the egg or
the color of the egg, *not the precision with which it is estimated against
the population of all cuckoo eggs.* What matters to mother reed-wren is
the *amount of the intruding egg's deviation from her birth babies' eggs
resting in the clutch.* Egg length and breadth in any clutch would not be
exactly the same to four significant digits, and one supposes it doesn't
need to be exactly so for evolutionary purposes. But apparently there is a
length and a breadth, in millimeters, that causes *too much* "inconvenience." The too long or (more to the point) too fat cuckoo egg gets "dismissed"—ejected—from the nest.

Whether or not ejection will occur is not *something that a calculation
of standard deviations can determine.*

Decades before Fisher, Pearson held the cuckoo notion that statistical
significance is capable of revealing scientifically important differences *from
the probability numbers themselves.* Though Pearson and Latter admit
that size is what matters *in theory,* they dropped that concern when they
got to the empirical section of Latter's article. It is not in fact the relevant
measure of size that Latter examined or evaluated when he made his calculations. Latter measured, under Pearson's editorial guidance, the ratio of
the signal to the noise, exclusively, and without regard to the inconvenience imposed on the foster mother. There are ways other than getting inside the mind of the victim to know what matters to her. For instance, one
could measure with some difficulty and sacrifice (but good science is difficult and sacrificial) the deviation in width of eggs of cuckoos that *fail* in

their perfidious scheme. But with Karl Pearson at his side, Latter looks at the world of birds with the sizeless stare of statistical significance.

On page 168 Latter publishes—in 1901, we note again—a prophetic summary table. Its final column is labeled "Significance Test (Ratio of difference to its probable error)." The "tests" are reported thus: "difference not significant (1.1)," "difference *significant* (3.71)", "difference not significant (1.25)" and so forth, by species of foster parent and for the variables length and breadth. It all sounds so very significant. But Latter's table is a very early example of the tabular presentation of individual tests of significance by variable that reports *only* "significance" and the lack of it, a practice, we have shown, now commonplace in all of the statistical sciences.

Pearson, we have noted, thought highly of young Gosset. He welcomed him to the Biometric Lab, and for the next three decades he published Gosset's most important scientific articles. A friendship evolved, reflected in forty surviving letters from Gosset to Pearson. But Pearson himself did not grasp the point of Gosset's findings on small samples. On September 17, 1912, Pearson wrote to Gosset, saying it made little difference whether the sum of squares was divided by n or the rigorously correct $(n - 1)$ "because only naughty brewers take n so small that the difference is not of the order of the probable error!"

After the Gosset year of miracles of 1906–7 Pearson would mention in passing small samples in classroom lectures to his students. But in practice he dismissed small samples on grounds that they are too "hard." Pearson wrote: "The knowledge that flows from small samples is, however, one of the hardest parts of our subject, and must be treated at a more advanced stage"—but the advanced stage never came.[1] Following K. P.'s death, his son Egon undertook a large survey of his father's life and work. He found that K. P. had not included Student's t in lecture notes or syllabi until 1921—fully sixteen years after Gosset had first mentioned to him the problem posed by small samples and its solution. K. P.'s eventual introduction of Student's t to the University College curriculum, Egon believed, was "perhaps in response to submissions from younger members of staff."[2]

Like scientists today in medical and economic and other sizeless sciences, Pearson mistook a large sample size for definite, substantive significance—evidence, as Hayek put it, of "wholes." But it was, as Hayek said, "just an illusion." Pearson's columns of sparkling asterisks, though quantitative in appearance and as appealing as is the simple truth of the sky, signified nothing.

20

Gosset:
The Fable of the Bee

Gosset came in to see me the other day. He is a very pleasant chap.
Not at all the autocrat of the t *table.*

YULE, N.D., QUOTED IN M. G. KENDALL 1952, 159

Given plenty of room bees rarely swarm.

GOSSET TO KARL PEARSON, APRIL 24, 1910
(FOLDER 704/7, PEARSON PAPERS, UCL)

Karl Pearson couldn't learn from Gosset. But Gosset couldn't learn much from Pearson, either. "I am bound to say that I did not learn very much from his lectures," he told Egon. "I never did from anyone's and my mathematics were inadequate for the task," he continued. "I had learnt what I knew about errors of observation from Airy" (i.e., from the astronomer G. B. Airy's *Theory of Errors of Observations* [1861]).

Gosset's lecture attendance, though, was good, and "Student" was a diligent note taker—until one day he came upon the idea of z (as t was at first called) for small samples, at which point his notes on Pearson's lectures go blank. Gosset thought his discovery "commercially important" on top of being scientifically correct, and he pursued it with all energy. He would later credit K. P. for "supplying the missing [mathematical] link in the probable-error-of-the-mean paper—a paper for which he [i.e., Karl Pearson] disclaimed any responsibility. I also learned from him how to eat seed cake," Gosset said, "for at 5 o'clock he would always come round with a cup of tea and either a slice of seed cake or a *petit-beurre* biscuit, and expect us to carry on."[1]

Fisher, unlike Pearson, did at first learn from Gosset. But his learning was of a selective and even monotypic kind and ultimately parasitic and manipulative in the style of a cuckoo. Gosset and Fisher became acquainted when the younger man and better mathematician wrote to Gosset at the Guinness Brewery in 1912. The letter and with it an enclosed technical note by Fisher concerned Gosset's 1908 article on what Fisher would later call "tests of significance" for small samples—Gosset's z-test.[2] Fisher greatly admired Gosset's article but pointed out a technical error: z should be divided by "degrees of freedom," not sample size, a correction Gosset agreed with and which caused z to be renamed, in 1922, Student's t.[3]

Gosset was impressed by the Cambridge boy's ammendment. Impressive but difficult-looking maths, he thought. Then he lost the note—in, he feared, one of the northern lakes! Gosset, who was a sailor and fisherman and general outdoorsman, worked by managed chaos. His scientific notebooks resemble grandmother's recipe books. One surviving notebook, for instance, contains a postcard from his brother Henry, a bill of receipt for sulphate of potash, a letter from Yule about some experiment they had previously discussed, and a formal memorandum from an Irish official—all scribbled over with Gosset's calculations and aphorisms and stuffed between the notebook covers. Mail was to Gosset just another stack of scratch paper. His office at the brewery, Egon Pearson has said, was a pulsing mess.[4]

Eventually, it seems, Fisher mailed Gosset another copy of the note. Gosset was grateful for it, but it was still "mathematics of the deepest dye," he told Karl Pearson. Gosset asked K. P. to publish Fisher's correction as a "note" in *Biometrika* and he did so—which marked both the beginning and the end of K. P.'s harmonious relationship with Fisher (E. S. Pearson 1968, 445–8). Meantime, Gosset extended his usual courtesy to Fisher. By 1918 Fisher's salutation "Dear Mr. Gosset" became in a letter the British-male-familiar "Dear Gosset," and so remained in the 150 surviving letters to follow.[5]

THE BEE WAS THE TEACHER

Gosset rarely depended on the test, even at the brewery (Ziliak 2008a). American hops preserved beer longer than Kent-raised hops, Gosset for example found—important to know when you're marketing a natural, unpasteurized stout such as Guinness, shipped worldwide. The shelf life of stout stored in a cask, he found, using hops imported from midwestern American farms, was three to seven days longer than one using hops from Kent.[6] But Gosset in this study and others often found z or t beside the point. "You want to be able to say 'if farmers [or whomever] in general do this [i.e., follow a certain experimental method] they will make money by it.'" A criterion of merely statistical significance could not satisfy such taste.

He "Hadn't a Jealous Bone in His Body"

Gosset appears by all accounts a good man. He never "fussed," said Launce McMullen, a friend and junior colleague at the brewery. "In personal relationships he was very kindly and tolerant and absolutely devoid of malice"—"unfortunately others were not always equal to this."[7] Another friend asked rhetorically, "Did you ever hear Gosset say an unkind thing about anyone?" and answered the question, "[Gosset] had an excuse for all the failings of other people, and how he enjoyed life—wet or fine—in bad days and good!" Deming, who was in some regards a Gosset-like character himself, thought Gosset "very humble, and of pleasing personality." Gosset's daughter Bertha told Egon, "I feel sure Dad must have been influenced by Dr. Spooner [the dean of New College, Oxford, Gosset's alma mater] because the two men had much in common, especially a deep integrity, wide interests, humility, and a capacity for taking infinite pains." Apparently no one, not even Fisher, could shake Gosset's "immovable foundation of niceness."[8]

Born in Canterbury, June 13, 1876, the eldest son of Col. Frederic Gosset and Agnes Sealy Gosset, William studied at Winchester College and New College, Oxford, where he obtained a first class in mathematical moderations in 1897 and in chemistry in 1899. The "first class" in mathematics he saw as mere ornament, no big shakes. In saying so he mocked neither mathematics nor his college (the number theorist and memoirist G. H. Hardy was a classmate) but his own abilities. In a letter to Karl Pearson Gosset confessed, "I don't feel at home in more than three dimensions."[9] He wrote to Fisher, "My mathematics stopped at Maths. Mods. at Oxford, consequently I have no facility therein."[10]

Gosset's "Original Small Sample Notebook," the one he used in 1906–7 to invent the t-test, contains only a few equations for the probable error of a mean, the probable error of a correlation coefficient, and the probable error itself. Formal equations and analytic proofs Gosset found to be secondary. By contrast, Gosset told Fisher, his friend Hardy did the reverse: he "always did scorn *applied* Mathematics except cricket averages."[11] The original notebook contains one proof in the style of mathematicians, though botched. Mostly the book is page after page of data collected by Gosset, his pioneering "Monte Carlo" samples of McConnell's 3,000 observation sample of heights and finger lengths, followed by columns of successful and unsuccessful arithmetic and doodles and more doodles (Ziliak 2008a). Only a few systematic plots are there—right-skewed and rectangular distributions, nonnormal distributions in small samples, it is important to note.[12]

In 1932 Gosset wrote a reply to Karl Pearson, who had late in life decided to tackle Gosset's small-sample test. But Gosset's reply, emphasizing his and Egon Pearson's economic and power interpretation of the test, "how we argue," was not what the aging positivist had hoped for:[13]

> Holly House
> Blackrock
> Co. Dublin
> 29.3.32

Dear Pearson

While it would not be reasonable for me to expect you to see eye to eye with me in the matter of Student's *z,* it does seem to me that *you do not quite realize just how we do use it in practice.* I have therefore ventured with a reply for your own perusal & not for publication—& enclose it herewith.

The method of argument—i.e. testing the hypothesis that *x* = ζ & so on is of course *that which E.S.P.* [Egon S. Pearson] *has been expounding in his recent papers & he has very much clarified my ideas as to how we argue....*[14]

But if we are mistaken the question should not be left as it is:—it is the universal practice to use correlated material where possible & if it is wrong it is important in the interests of experimental work that it should be proved to be wrong. Could Miss David's experimental sampling be carried further?

> Yours v. sincerely
>
> W S Gosset

The great feminist statistician Florence Nightingale David (the "Miss David" to whom Gosset refers), before moving to the University of California-Riverside, had taught for many years at University College London. She knew well the leading men of English-language statistics. "Went fly fishing with Gosset," she said. "A nice man. Went to Fisher's seminars with Cochran and that gang. Endured K. P. . . . Spent three years with Neyman. Then I was on Egon Pearson's faculty for years."

> Fisher was very vague [she said]. Karl Pearson was vague. Egon Pearson vague. Neyman vague. Fisher and Neyman were fiery. Silly! Egon Pearson was on the outside. They were all jealous of one another, afraid somebody would get ahead. *Gosset didn't have a jealous bone in his body. He asked the question* [about power and alternative hypotheses]. Egon Pearson to a certain extent rephrased the question which Gosset had asked in statistical parlance. Neyman solved the problem mathematically. (quoted in Reid 1982, 132–3; italics supplied)

A WOODY GUTHRIE OF MATHEMATICAL STATISTICS

A passionate, even somewhat loony, hobbyist, Gosset was in other regards, too, his own man. He repaired model and full-sized fishing boats with penknives and designed an original fishing boat featured in *Field* magazine—he fitted the boat with opposing rudders, allowing it to steady in the water without anchoring. He hiked the Dublin Mountains, took long bike rides, made animal noises, and crossbred and harvested edible berries, breeding what he called the "jamberry," which Fisher found delicious. He went on fishing trips with his wife and would take his children to intellectual meetings he thought they'd like. Though not a churchgoer, he was "careful not to say anything that would undermine" the children's "faith." He golfed with strange, antiquated clubs (McMullen 1939, 209). "He was a sound though not spectacular shot," a friend recalled, "and was well above the average on skates" (209). He tended apples, played the pennywhistle, listened to Beethoven on the phonograph, mentored young scientists, kept lifelong correspondents, and frequently visited his father in rural England.[15]

Sweet came naturally to Gosset. In an article in *Biometrika* he explained the meaning of kurtosis with his own hand-drawn figures of a platypus and two kangaroos (Student 1927, 160; see fig. 20.1). He fancied things "pretty" and "natty in Γ-functions."[16] The man with the "immovable foundation of niceness" was thoughtful of others, shipping packages of wasps and bees to K. P., for instance, to keep his friend abreast of Edgeworth's thinking. Once he sent by mail to Fisher an unusual chicken foot—it was born in Gosset's yard with one too many toes—thinking

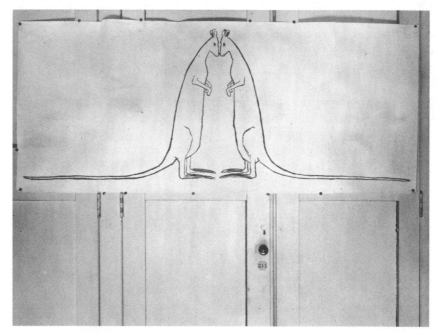

Fig. 20.1. "Real Error" and Gosset's *Memoria Technica*. An intuitive and playful man, Gosset expressed his scientific ideas with the help of figure drawings. With a drawing of platypus and kangaroos (above), he tried to persuade Egon Pearson that the magnitude of "real error" can only be discovered through "repetition" on other samples, particularly when a single sample on offer is nonnormally distributed, such as platykurtic (a sample distribution with shorter than normal tails) or leptokurtic (longer than normal tails). "I myself bear in mind the meaning of the words," Gosset said, "by the above *memoria technica*." (Student 1927, 160. Drawing courtesy of University College London, Special Collections Library.)

Fisher would be interested in it. At scientific meetings Gosset mostly listened. On his bicycle he wore rugged denim and tall leather boots, and carried a rucksack on his back, where he kept "Baby Triumphator," his calculating machine. Charming, rustic, humble, and mysterious, Gosset was a very Woody Guthrie of mathematical statistics. His "machine" did not "kill fascists," as Guthrie would say about his guitar, but he did everything he could to humanize the London statisticians.

"STUDENT"

Especially Gosset's humility is unusual, equaled in the front ranks of statistical science perhaps only by that of Harold Jeffreys, announced by the pseudonym, "Student."[17] Why the pseudonym? Guinness, like most companies at the time, made anonymity of authorship a condition. *Biometrika* and other journals were thus authored by "Sophister" (G. F. E. Story, a Gosset protege), "Mathetes" (Edward Somerfield, a Gosset assistant), and our "Student." The name comes we think from the maker of one of Gosset's early, prepublication notebooks used in the crucial years 1905–7, which is labeled on the cover "*The* Student's *Science Notebook*, Eason and Son, Ltd., Dublin and Belfast."[18] Just a student, then, a little worker bee.

Student's real identity was known only to colleagues of his immediate acquaintance. Although Student was by the 1930s world famous in agronomy, the design of experiments, and mathematical and applied statistics, the world did not know who stood behind the pseudonym. Gosset did not openly reveal his identity until 1936, when he tried at a meeting of the Royal Statistical Society to check a blustering Fisher, who was making again a pitch for his "randomized," antieconomic design of experiments (Gosset 1936, 115; cf. Jeffreys 1939b).

In the ego-soaked societies of Karl Pearson and Ronald Fisher, the humility of Gosset's anonymity is stunning. "Student" itself contrasts with Somerfield's proud "Mathetes"—a "student," too, but in Greek. Egon wrote (1939, 248–49), "[A]ll who have known him will agree that he possessed almost more of the characteristics of the perfect statistician than any man of his time . . . quiet and unassuming, who worked not for the making of personal reputation, but because he felt a job wanted doing and was therefore worth doing well."

Fisher:
The Fable of the Wasp

It will be seen then that the difference between Prof. Fisher and my-
self is not a matter of mathematics—heaven forbid—but of opinion.
STUDENT 1938, 367

By how much *may we expect the yield of variety* B *to exceed that*
of variety A *if they were sown alternatively on the same soil in the*
same season?
STUDENT 1926, 126

If Gosset was the Bee, his difficult friend Fisher was the Wasp. Gosset pa-
tiently tried for a quarter century to teach Fisher about human relations,
such as the importance of being kind and telling the truth and practicing
humility and giving credit to other scientists and being accurate about history. He tried to teach the Wasp about the Wasp's own β-self, hoping Fisher
would get around then to analyzing the *how much* of his β-coefficients.

But this Wasp was not an apt student in matters of scientific ethics. He
was not inclined to give anyone beyond himself scientific credit. Fisher
died without acknowledging, for example, Edgeworth's 1908 original insights on maximum likelihood, misleading historians of statistics for half
a century.[1] The Bee nudged him along in those 150 letters (Gosset [posthu-
mous] 1962). But the Wasp, obsessed as he was with his own α-self,
wouldn't listen, even after he made significant errors in his sometimes brilliant but uneven and difficult youth.[2] The Wasp, in the end, never did
make time for his β-self.

The Bee did his best. He tried, for example to explain to the Wasp how
to design and evaluate experiments *economically,* tasks at which the Bee

VOLUME VI   MARCH, 1908   No. 1

BIOMETRIKA.

THE PROBABLE ERROR OF A MEAN.

By STUDENT.

Introduction.

ANY experiment may be regarded as forming an individual of a "population" of experiments which might be performed under the same conditions. A series of experiments is a sample drawn from this population.

Now any series of experiments is only of value in so far as it enables us to form a judgment as to the statistical constants of the population to which the experiments belong. In a great number of cases the question finally turns on the value of a mean, either directly, or as the mean difference between the two quantities.

If the number of experiments be very large, we may have precise information as to the value of the mean, but if our sample be small, we have two sources of uncertainty:—(1) owing to the "error of random sampling" the mean of our series of experiments deviates more or less widely from the mean of the population, and (2) the sample is not sufficiently large to determine what is the law of distribution of individuals. It is usual, however, to assume a normal distribution, because, in a very large number of cases, this gives an approximation so close that a small sample will give no real information as to the manner in which the population deviates from normality: since some law of distribution must be assumed it is better to work with a curve whose area and ordinates are tabled, and whose properties are well known. This assumption is accordingly made in the present paper, so that its conclusions are not strictly applicable to populations known not to be normally distributed; yet it appears probable that the deviation from normality must be very extreme to lead to serious error. We are concerned here solely with the first of these two sources of uncertainty.

The usual method of determining the probability that the mean of the population lies within a given distance of the mean of the sample, is to assume a normal distribution about the mean of the sample with a standard deviation equal to $s/\sqrt{n}$, where s is the standard deviation of the sample, and to use the tables of the probability integral.

Biometrika VI   1

Page 1 of Student 1908a. Karl Pearson encouraged Gosset to publish "The Probable Error of a Mean," though Pearson did not see much value in it. The article went unnoticed until 1912, when Fisher wrote to Gosset to tell him about "degrees of freedom." Fisher saw in Gosset's articles of 1908 (1908a, 1908b) a revolution in the life and social sciences and for Fisher himself a major reputation in statistics, and produced them both. (Reproduction courtesy of Oxford University Press and *Biometrika*.)

had tried, failed, practiced, and improved upon for decades before the Wasp took his first post at Rothamsted. Gosset actually ran experiments on things that mattered to a bottom line, your uncle's favorite beer, for example, and at the time the world's largest brewer (Ziliak 2008a). Still the Wasp wouldn't listen.

In July 1923 Fisher, who had recently published results on one of his first experiments, wrote sharply to Gosset from Rothamsted—a post Gosset helped him to secure—asking Gosset how *he* would have designed the experiment.[3] "How would I have designed the exp[eriment]?" Gosset replied to his adopted student. "Well at the risk of giving you too many 'glimpses of the obvious,'" Gosset, an experimentalist with by then two decades of experience, wrote, "I will expand on the subject: you have brought it on yourself! The principles of large scale experiments are four," he explained to Fisher in a reply dated July 30. "There must be essential similarity to ordinary practice. . . . Experiments must be so arranged as to obtain the maximum possible correlation [*not* the maximum possible statistical significance] between figures which are to be compared [like Leamer and other oomph-ful scientists, Gosset thought in terms of upper and lower bound estimates, best and worst case scenarios]. . . . Repetitions should be so arranged as to have the minimum possible correlation between repetitions (or the highest possible negative correlation). . . . There should be economy of effort [net pecuniary advantage in the 1905 sense]."[4] Fisher shrugged. The economic approach to the design of experiments was too difficult. He never did try Gosset's way.[5]

Fisher Was "Insensitive to Fellow Humans"

Gosset's methods sweetened the Guinness bottom line. But the Wasp, a Guinness drinker, didn't care. Though an undoubted genius, a major figure of science in the twentieth century, Fisher was considered by friends and associates to be a blusterer. And yet he coveted the insights of the Bee, the anonymous Student, who was in his own way a genius. And the Bee's lessons were packaged in a sweet scolding, perhaps irresistible. "I have come across the July J. A. S. [*Journal of Agricultural Science*] and read your paper," the Bee wrote to the Wasp. "I fear that some people may be misled into thinking that because you have found no [statistically] significant difference in the response of different varieties to manures that there isn't any. The experiment seems to me to be quite badly planned, you should give them a hand in it; you probably do now."[6]

In 1925 Fisher published *Statistical Methods for Research Workers*—one of the two most influential books of statistics in the twentieth century, the other being *The Design of Experiments* (1935), also by Fisher. *Statistical Methods* covered a lot of territory and was in many regards an original effort (unlike the thousands of imitative textbooks that have followed). Short on proofs and heavy on concepts and examples, its influence on the scientization of economics, agriculture, biology, medicine, psychology, law, and other fields cannot be underestimated.

When *Statistical Methods* came out, Fisher had only recently begun to develop his own ideas about experimentation. So, he gave only a few sections of the first great book to it. And though much of his education on testing, estimation, and experimentation had come from Gosset, he set aside Gosset's economic principles without mentioning them—or him.

Gosset noticed. Seeing a set of prepublication page proofs (which Fisher had sent to him) with unusually few corrections to the experimental sections, Gosset suggested to Fisher that the lack of corrections might be "possibly because of his [i.e., Fisher's inexperience at that stage in actual experiments, and so] understanding less of the matter."[7] On the matter of Fisher's "random block" approach to experimental design, Gosset wrote, "I don't agree with your controlled randomness. . . . You would want a large lunatic asylum for the operators who are apt to make mistakes enough even at present. . . . If you say anything about Student in your preface you should I think make a note of his disagreement with the practical part of the thing." Significantly, Fisher did not say anything of the kind in the preface about Student or anywhere else in his writings.

Among the first reviews of Fisher's *Statistical Methods for Research Workers* (first and second editions) were those of Egon Pearson. He reviewed the first edition in *Science Progress* and the second edition, in 1928, in *Nature.* Fisher responded angrily to Egon's review in *Nature,* and "there followed a chain of twenty-eight letters [some of them published] extending over four months of 1929" (E. Pearson 1990, 95). Gosset the Bee was asked to intervene, in his usual counselorlike role, and made a special visit to Gower Street to visit Egon.[8] Gosset lent a sympathetic ear to the Wasp, too, but he was particularly sensitive to the wounds of Egon, whom Fisher persistently belittled. Fisher's position was that Fisher could never be wrong about a scientific position and especially not in public. Pearson *fils* was, like his father, interested in problems of estimation and testing when the data deviate considerably from normality, an interest he seems to have got not from his father (as one would think) but from Gosset's biometric research of 1904–8 and his letters of 1926. Fisher replied that everything could be seen as "normal" and implied that Student's *t* tables were anyway "exact," even for very small, allegedly nonnormal samples, an assertion that Gosset, Egon, and Gosset's assistant Somerfield had long contradicted.

Fisher wrote in a letter on June 27 , 1929, to Gosset.

> Of course I disagree with your last letter entirely. What has Somerfield got to do with it? As I understand it there is a α-Somerfield who would not give two hoots for normality, and β-Somerfield who is shocked at his ignorance and indifference. You think β-Somerfield is the wiser man, so

does he; in the absence of evidence I cannot see any ground for judging between them. But you are not content with condemning α- as the villain; it appears that I am responsible for his villainy, which I could not be even if I had nursed him through his teethings. (Fisher to Gosset, Letter 103, in Gosset 1962)

Or so α-Fisher would say.

In *The Design of Experiments,* the other great book, published ten years after *Statistical Methods,* Fisher again did not acknowledge Student, *or Gosset—at all.* Experimental design, unlike significance testing, was something Gosset really valued. He taught Fisher the basics of both techniques.[9] But Student's name is mentioned in *Design* only in reference to Student's *t*—the one invention of Gosset's, we feel we must repeat, that Gosset considered dangerous to human intelligence, at best "valueless" in Fisher's hands.[10] In his obituary on Gosset, Fisher said that Gosset was a "failure" (Fisher 1939, 7) in his economic approach to the design of experiments. He could never say why or wouldn't. Others, such as Egon, Neyman, Jeffreys, Beaven, and the world's largest producer of beer appear to have disagreed with Fisher.[11] Gosset's innovation of the half-drill strip method, for example—which he with Beaven proposed as an alternative to Latin squares, chequerboards, and (later) Fisher's "artificial randomization" in the layout of field experiments—selected and cultivated a barley variety that Beaven himself, the chief purchaser of barley for Guinness, considered to be superior in quality and yield to the thousands of extant varieties, America to New Zealand, he had seen in "50 years of observation and experiment."[12] Rather than mention anything of the sort, rather than mentioning his decade-long apprenticeship in Gosset's theory of economic significance and experimental design, he called the *t*-test of statistical significance Gosset's "great" contribution to science (as interpreted, of course, by Mr. Fisher, *not* by the late Mr. Gosset). He was being nice to Gosset, he probably thought, writing him into the annals as one of the "greats," as undoubtedly he was.

But if that was being nice, the Wasp wasn't nearly so nice to other rivals. And Fisher decided to have a great many rivals. Gosset wasn't in the line of fire, and it's interesting to think about why. Fisher never spoke of Gosset while alive as rival or competitor or longtime mentor, though to Fisher, privately speaking, Gosset was all of those things. Fisher's failure to publicly acknowledge the alternative ideas of his generous and ingenious friend led the way to active disrespect and exploitation. The great and amiable statistician Jerzy Neyman, to speak of one acknowledged rival, was definitely in the line of fire. By contrast Neyman suffered numerous printed insults from Fisher, and Fisher can be shown to have blocked appointments for him.[13] In 1955, for example, he raged against Neyman as though he, not Gosset or, before Gosset, Gauss, or before Gauss, Laplace, had introduced "cost functions" into testing and estimation procedures (Neyman 1956, 293–94). Fisher detested the idea of assessing the gains and losses from estimation in any currency. And since Neyman's name was increasingly identified with cost functions (and good jobs), he detested him. As for Egon Pearson, an equal and original partner in the Neyman-Pearson approach, Fisher was dismissive. Egon, despite a kind manner and strikingly original contributions, Fisher treated as irrelevant, an insult directed more at the father than the son.

When the elder Pearson retired from University College in 1933 his duties in statistics and eugenics were split in half. The division was intended to solve both a social and intellectual problem. In the new arrangement, Egon, at that time a reader, not a professor, was to run the Department of Statistics, and Fisher, already a professor, was to run the Department of Eugenics as the new Galton Professor of Eugenics (E. Pearson 1938, 231). Neyman, in Poland and still desperate, heard the news and wrote immediately to Fisher: "[N]ew people will be needed . . . please consider whether I can be of any use." Replied Fisher: "I think Egon Pearson is designated as Reader in Statistics. This arrangement will be much laughed at, but it will be rather a poor joke, I fancy, for both Pearson and myself. I shall not lecture on Statistics . . . so that my lectures will not be troubled by students who cannot see through a wire fence. I wish I had a fine place for you [in the Galton Laboratory]."[14] No one (including a now physically ailing Karl, the outgoing Galton Professor) liked the arrangement. But Egon was in one respect content: as Chair of the new Department of Statistics and Director of the Biometrics Laboratory, his first success was to rescue Neyman from Poland and bring him back to University College as a temporary reader in *his* Department.

By 1935, in other words, Egon, Fisher, and Neyman were working in the same building on Gower Street. (Gosset was nearby, too, in Park Royal, but he was rarely seen on Gower Street. He was busy operating at Park Royal the new Guinness brewery.) Without Gosset the Bee to smooth things over with Neyman and Pearson, Fisher the Wasp stung repeatedly. On one occasion, for example, Fisher stopped by Neyman's office on his way to a crucial faculty meeting at which were to be considered two proposals Fisher fiercely opposed: the promotion from reader to professor of wire-fence

Pearson and, worse, a permanency for Neyman's readership, for that Pole, that substantivist, that radical. The Wasp popped his head around Neyman's open door and declared loudly that Neyman should leave England entirely, "for California" or some such backwater, because Fisher would block all honors for him—unless, indeed, he, meaning Neyman, would promise to teach only out of Fisher's *Statistical Methods for Research Workers* (Reid 1978, 126). When Neyman refused, Fisher strode out, slamming the door. The man was more than a little envious of the power now enjoyed by Egon and Neyman. At the meeting he did not prevail.

"The Trouble of Marking Individuals"

Fisher's second of five daughters, the historian Joan Fisher Box—married to the Wisconsin statistician the distinguished George E. P. Box (b. 1919)—writes that envy and bad judgment in personal matters was a family trait. Her father, she said, loved his family but "ineptly" (Box 1978, 10–11).[15] On one occasion Fisher insisted that it was time for his young niece, Kestrel, who lived with him, to learn how to survive on her own (49). So he sent her to cross a wood, about a mile long, to deliver a homemade cheese to the village on the other side. Hours later Fisher's sister-in-law (a cousin, it turns out, like his wife, of the Dublin Guinness family) set out in the wood, too, against Fisher's will, to search for her missing daughter (the two families lived communally and, they said, "eugenically," in a large house in the wood [43]). She found her daughter wandering alone—dirty, shaken, and badly confused. The child sent into the woods by Fisher was three years old. When Fisher accepted the Arthur Balfour Chair of Genetics at Cambridge University in 1943, the peak of his academic career, he left his wife and eight children behind. The new professor installed in his house in Cambridge not his family but hundreds of laboratory mice, on whom he could work in genetic experiments. Divorce soon followed.[16]

Fisher's mother, writes Box, "passed on to her son her own emotional inadequacy. . . . Their communication at an emotional level . . . was stilted and unconvincing, and they were curiously oblivious of the feelings of others" (1978, 10). As a boy, afflicted with asthma and poor eyesight, Ronald leaned on her. She could in one sense help him immensely: the "intellectual channels were wide open." When his mother read excitedly from books of astronomy or history there "he sat enthralled beside her," on the floor, "on a velvet cushion." But relatives "thought [the mother] poisonous. In their occasional meetings they found her selfish, indolent, and arrogant, with ideas above her station." Her "reputation was shared by her daughters. . . . [Like Ronald's sister Sybil] she offered no lifelines at an emotional level. . . . It is no wonder that she inspired both devotion and bitter resentment" (9).

An ethical life in science seems to require an emotional life outside of it. The *Autobiography* of John Stuart Mill and the "non-statistical character" of William Sealy Gosset, to take two examples, suggest such a hypothesis (E. Pearson 1939, 211). Fisher "grew up without developing a sensitivity to the ordinary humanity of his fellows," Box says. "He was unaware of the effects of his own behavior" (1978, 10). For instance, Fisher once came into his genetics lab to find that an assistant, a young woman, had not divined some instruction that he had failed to transmit. With his bare hand Fisher picked up a laboratory mouse and began to verbally abuse the lab assistant. In his rage he crushed the mouse to death. The young woman in front of him grew pale and stiff; she was mortified. Fisher, without a word, left the room. He returned momentarily, showing his face at the door, grinning in a Cheshire cat kind of way, and left again, without further apology or explanation. On another occasion, in the early 1930s, Neyman had used little painted blocks in a seminar to demonstrate that Fisher's preference for "randomized blocks" in experimental design was unfounded (Fisher 1935, chaps. 2, 4). Egon, Neyman's partner in the crime, was in attendance. Neyman and Pearson were trying to prove mathematically what Gosset for twenty years had been proving experimentally and verbally: Gosset's "balanced designs" and "half-drill strips," not Fisher's "randomized blocks," are the real money and oomph makers.[17] Fisher was enraged and, later that evening, back at the Statistics Department, relieved his anger by tipping over their cabinet full of blocks onto the hallway floor.

"I am surely not alone in having suspected that some of Fisher's major views were adopted simply to avoid agreeing with his opponents," wrote Neyman.[18] And Zellner, who knew personally all of these men, wrote, "As regards Fisher and his impact, he was indeed a very influential and forceful person. The joke is that when he had a disagreement with Jeffreys, given his forceful manner, he won the argument whether he was right or wrong" (2004b). The joke, Zellner agrees, goes beyond the point of diminishing returns.

Raymond Birge, a professor of physics at the backwater of the University of California-Berkeley, was distressed to learn in 1936 that the

Wasp would spend a whole month in residence there. Birge's friend Deming, who had just hosted Fisher at the U.S. Department of Agriculture, warned Birge of Fisher's next stop. Birge tried to make the best of it, arranging for meetings and outings and time for Fisher to be alone. "From Birge's point of view, however, Fisher's visit to Berkeley was a great disappointment," an understatement matching Deming's "disappointment" in Washington (Reid 1978, 143). Birge wrote a long memorandum detailing Fisher's asocial behavior. He recorded how Fisher rarely met with faculty or students. He did not keep his office hours. He skipped arranged dinners, even a special one in his honor. Birge wrote to Deming: "As you [yourself] have written, he is glad to discuss . . . things early in the morning or late at night. *But* he is *not* glad or even willing to have others work on the purely theoretical aspects of his work. He expects others to accept his discoveries without even questioning them. He does not admit that anything he ever said or wrote was wrong. But he goes much further than that. He does not admit even that the *way* he said anything or the nomenclature he used could be improved in any way" (quoted in Reid, 144). Birge told Deming that Fisher was the most conceited man he ever met— in physics, he added, a stiff competition (perhaps Professor Birge did not know any economists). J. Robert Oppenheimer, who was at the time a member of the physics faculty at Berkeley, said, "I took one look at him and decided I did not want to meet him."

One can hear the imperious, arbitrary man declaring in successive years the truth of his 5 percent philosophy.

> The value for which P = .05, or 1 in 20, is 1.96 or nearly 2; *it is convenient to take this point as a limit in judging* whether a deviation is to be considered significant or not. Deviations exceeding twice the standard deviation are thus formally regarded as significant. Using this criterion we should be led to follow up a false indication only one in 22 trials, even if the statistics were the only guide available. (Fisher 1925a, 42; italics supplied)

> It is convenient to draw the line at about the level at which we can say: "*Either there is something in the treatment, or a coincidence has occurred* such as does not occur more than once in twenty trials." . . . If one in twenty does not seem high enough odds, we may, if we prefer it, draw the line at one in fifty (the 2 per cent point), or one in a hundred (the 1 per cent point). Personally, the writer prefers to set a low standard of significance at the 5 per cent point, and *ignore entirely all results which fail to reach this level.* A scientific fact should be regarded as experi-

mentally established only if a properly designed experiment rarely fails to give this level of significance. (Fisher 1926b, 504; italics supplied)

It is usual and convenient for experimenters to take 5 per cent. as a standard level of significance, in the sense that they are prepared to *ignore all results which fail to reach this standard,* and, by this means, *to eliminate from further discussion* the greater part of the fluctuations which chance causes have introduced into their experimental results. (Fisher 1935, 13; italics supplied)

A null hypothesis may, indeed, contain arbitrary elements, and in more complicated cases often does so: as, for example, if it should assert that the death-rates of two groups of animals are equal, without specifying what these death-rates actually are. In such cases it is evidently the equality rather than any particular values of the death-rates that the experiment is designed to test, and possibly to disprove. (Fisher 1935, quoted in Savage 1976, 471–72; italics supplied)

Though recognizable as a psychological condition of reluctance, or resistance to the acceptance of a proposition, *the feeling induced by a test of significance has an objective basis in that the probability statement on which it is based is a fact communicable to and verifiable by, other rational minds.* The level of significance in such cases fulfils the conditions of a measure of the rational grounds for the disbelief it engenders. (Fisher 1956, 43; italics supplied)

When decision is needed it *is* the business of inductive inference to evaluate the *nature* and *extent* of uncertainty with which the decision is encumbered. . . . We aim, in fact, at methods of inference which should be *equally convincing to all rational minds, irrespective of any intentions they may have in utilizing knowledge inferred.* (Fisher 1955, 75; italics on isolated words in original; italics in the last sentence supplied)

Finally, in inductive inference we introduce no cost functions for faulty judgments, for it is recognized in scientific research that the attainment of, or failure to attain to, a particular scientific advance this year rather than later, has consequences, both to the research programme, and to advantageous applications of scientific knowledge, which cannot be foreseen. In fact, scientific research is not geared to maximize the profits of any particular organization, but is rather an attempt to improve public knowledge undertaken as an act of faith to the effect that, as more becomes known, or more surely known, the intelligent pursuit of a great variety of aims, by a great variety of men, and groups of men, will be facilitated. *We make no attempt to evaluate these consequences, and do not assume that they are capable of evaluation in any currency.* (Fisher 1955, 75; italics supplied)

To evaluate size-matters/how-much would have forced Fisher to listen to and cooperate with others. Determining whether something matters to people depends on actually listening to people, as a heart surgeon listens to a radiologist, as a beer brewer listens to a customer. Admitting that size matters would have required Fisher to admit that regression coefficients "are capable of evaluation in any currency." It would have put him in the unhappy position of having to communicate with others about the meaning of his findings. This, we have shown, he would not do. And so Fisher substituted a metaphysical question of Whether, attained by calculations performed in isolation, for the difficult but scientific questions of How Much and Who Cares, which are answerable only in conversation with other scientists. In Fisher's own journal of genetics, *Heredity*, nary a significance test was to be found during his tenure as editor. Odd. His followers have, by contrast, indulged his sizeless test.[19]

Fisher himself indulged it in other outlets. In his own statistical work he rarely mentioned magnitudes—such as the size and meaning of correlation coefficients. In 1931 Gosset wrote to Karl Pearson, explaining again his friend's adoring but incorrect use of Gosset's test:[20]

St. James's Gate,
 Dublin
  14th July, 1931

Professor Karl Pearson, F.R.S.
 Galton Eugenics Laboratory,
  University College,
   London, W.C.1

Dear Pearson,

 I am enclosing a note on the use of z in testing the significance of the average difference between correlated variables. I hope you will see your way to put it in [*Biometrika*] *as I have always attached considerable importance to arranging matters so that the correlation should be as high as possible*. In the case of agricultural experiments, *it has been my chief criticism of Fisher that he does not take all possible steps in this direction*.

 I am at present engaged in a criticism of an experiment on giving school children a ration of milk.

 This was carried out on a very large scale at an expenditure of some £7,500 in Lanarkshire. Five thousand children having been given ¾ pint raw milk per day for four months, five thousand pasteurized milk and ten thousand no milk at all.

 By some mismanagement they chose heavier and taller children for

the controls and I imagine that this was due to the teachers who made
the selection taking pity on the less well nourished children and arranging that they should get the milk [they didn't choose a random sample;
therefore, an ordinary test of significance was going to contain what we
know call a "sample selection bias" of unknown magnitude]. In any case
there seem to be some other ways in which the experiment is open to
criticism.

You may have seen in "Nature" a note by Fisher and a man at Reading. The original authors had decided that there was nothing to show
any difference between pure and pasteurized milk; *Fisher showed that the
difference was* [statistically] *significant; my work will, I think, show that
the selection which took place prevents our drawing any conclusion in
the matter* . . .

Yours very sincerely,

W S Gosset

Gosset, we have seen, is not the only colleague of Fisher's who has made
our point. Egon and Yates and Deming and Wald and Savage did, too. In
his "Note on an Article by Sir Ronald Fisher," Neyman was especially
clear: "Another item worth mentioning in Sir Ronald's [1955] section on
errors of the second kind is the passage . . . [in Fisher] 'it is a fallacy so well
known as to be a *standard* example, to conclude from a test of significance that [if $t < 1.96$] the null hypothesis is thereby established.'" (More
despair, you can see, by a man losing status with the highbrows.) But, as
Neyman remarks, "Although no names are mentioned [by Fisher], the
context suggests that the fallacy in question is committed by the same
people who are guilty of considering alternative hypotheses and the power
of tests. Whereas Sir Ronald abstains from quoting any specific instance,
it is easy to quote those in which he himself, and his followers, acted pre-
cipitately and advised others to do likewise, when a test failed to detect a
significant effect. A case in point is the design of the factorial experiment"
(1956, 290; italics in original). A second "case in point is Fisher's famous
experiment concerned with the Lady Tasting Tea" (290, n. 3). A third case
in point is Fisher's so-called exposure of fraud in Mendel's peas (Fisher
1936; Mayr 1982, 719–20).

Some will see Fisher's own statistical practice in a different light. But
the seeds of the Standard Error were incontrovertibly sown in all of the
sizeless sciences by the great man of Rothamsted. Scientists, Fisher pro-
claimed against the common sense of a beer brewer, should "introduce
no cost functions" to a test of significance. They should not care about the

size of experimentally determined death rates or the loss of human life in treatment and control groups, merely whether or not the death rates are equal—"without specifying what these death-rates actually are." "Don't try to say what you mean by 'children and the right amount of nutrition,'" he thundered. "Just use my null procedure to 'test' whether or not the nutrition rates are equal across treatment and control groups." Scientists, Fisher said, should "not assume" their research is "capable of evaluation." They must not work to "maximize profit," he said in 1955, only for "faith"—a secular faith, he means, in the possibility that another mechanically calculated output of p-values by themselves could contribute to scientific progress. The scientist should not worry, as Gosset worries in the letter to Karl Pearson, and as Ernst Mayr worries in his comments on Fisher's (1936) tests of chi-square on Mendel's peas, whether their samples are random: just test, test, test, *as if* random.[21] A 5 percent level of Type I error is, when "formally" considered, says Fisher, the final judge of Science.

Here's a scientist
who sank the world with a *t*
5 percent per cup.

How the Wasp Stung the Bee
and Took over Some Sciences

How long does a wasp take in loading herself with sweets?
EDGEWORTH 1896, 358

In 1922 Fisher wrote to Gosset, soliciting Gosset's new table of *t*. Fisher
was eager to put the updated table in the first edition of his *Statistical
Methods for Research Workers*. They would call Gosset's tables "the table
of 'Student's' *t*," in celebration of Fisher's correction to Gosset's 1908
table of *z*. Gosset, as always, wanted to help. He was unusually busy with
work at the brewery. Baby Triumphator, his calculator, he told Fisher, was
needed for overtime work.[1] Calculating *t* values at various levels of significance as *N* goes from small to large was difficult manual labor. Gosset's own assistant, E. M. Somerfield (aka "Mathetes"), himself a Fisher
student and an accomplished statistician, was not strong enough to turn
the handle.[2] Fisher would have to wait.

Fisher's tone in reply was urgent. So Gosset put Somerfield onto the
task of preparing the index to Fisher's book. That freed some time for Gosset—who was planning himself to help with the index to the anti-Gosset
book—to turn the crank on Baby Triumphator. When he sent the new
t-values to Fisher, in September 1922, Gosset exclaimed with deficient fore-
sight, "[y]ou are the only man that's ever likely to use them!"[3] The men discovered some errors in the 1922 version and went back to work. But Fisher
then asked Gosset if he could "quote" the completed table in a *Biometrika*
article Fisher contemplated. "Dear Fisher," Gosset replied in July of 1923,
"I expect to finish it sometime next winter. I should say that it is certainly
in course of preparation. As to 'quoting' the table in *Biometrika* it depends

just what you mean by quoting. . . . I don't think, if I were Editor, that I would allow much more than a reference!"[4] For over a year they worked out the bugs in the table, Gosset doing most of the work.

THE STING

In Student's name the original table of z had first been copyrighted, in 1908, in Pearson's *Biometrika*. A second version was published by Student in Karl Pearson's *Tables for Biometricians and Statisticians* (1914). A third, fuller version of z (with more n) was copyrighted by Student in 1917, also in the Pearson-edited journal (Student 1908a, 1917). Fisher in the early 1920s published "small sample" results in the Italian journal *Metron*, edited by Corrado Gini, first in 1921 and a second time in 1924 (Fisher 1950, 1.2a). By 1925 the Wasp had cocked his stinger.

Fisher published two articles in the December 1925 issue of *Metron*, "Applications of 'Student's' Distribution" and "Expansion of 'Student's' Integral in Powers of n^{-1}" (1925b, 1925c). Sandwiched between Fisher's two articles was a much shorter article (a little over three pages long), by Student himself, "New Tables for Testing the Significance of Observations" (1925). Student wrote, "The present Tables have . . . at Mr. Fisher's suggestion been constructed with argument $t = z \sqrt{n}$ where n is now one less than the number in the sample, which we may call n'" (106). The balance of Gosset's article explains in detail how Gosset calculated the t-tables. Fisher's first article, at fourteen pages long—he apologized in a letter to Gosset for excess length but went on to say that it should be longer yet—is by contrast rhetorically complicated and didactic, as Fisher was. His "Applications" demonstrates in n-dimensional Euclidean space what remains to Gosset "partly intuitive" (Fisher 1925b, 92)—the "exactitude of 'Student's' distribution for normal samples" (92). He then illustrates "significance of differences between means"—showing how "'Student's' distribution affords the solution to a variety of problems beyond that for which it was originally prepared" (94–96). He goes on to show—and this was truly novel, another Gosset-inspired idea—"the second class of tests for which 'Student's' distribution provides an exact solution, . . . testing the significance of the large class of statistics known as *regression coefficients*" (96; Lehmann 1999; italics supplied).

Unsurprisingly significance rules: "The multiple correlation must be judged significant," Fisher declares, "only if the value of P obtained is too small to allow us to admit the hypothesis that the dependent variate is really uncorrelated with the independent variates" (104). "Only if," he said. The value is to be admitted "only if"—as he wrote simultaneously in his Oliver and Boyd book of 1925—it meets his Rule of Two. The novelty and promise of the application of Student's distribution to regression coefficients, the freshness of the "completed" and "corrected" tables, and the confident scientific tone maintained throughout were probably important for its great influence on Cowles Commission econometrics and beyond.

But the sting in *Metron* is that the three tables of *t*, invented and then calculated nearly entirely by Gosset, are attached to *Fisher's* second article (Fisher 1925c, 113–20), as if Fisher was the inventor and calculator. In the same year the first edition of *Statistical Methods for Research Workers* (Edinburgh: Oliver and Boyd) came out. For it Fisher himself *copyrighted again Gosset's tables in his own name.* Fisher had made the crucial substitution that consigned Gosset to academic obscurity. Gosset, in his gentle way, expressed mild annoyance.

The *Metron* papers were intended originally for *Biometrika,* which Karl Pearson still edited. On June 12, 1925, six months before the Gini-edited *Metron* publication, Gosset wrote to Fisher, "K. P. is very anxious to publish your note about the use of the table, but doesn't like the binomial approximation which he considers requires a proof of convergence." Fisher wasn't moved and refused to do the proof. Fisher would do nothing to please the man who ten years prior had criticized his work in public—even if he had given him his career. (Fisher's betrayal of one friend after another reminds us of the psychiatric diagnosis "borderline personality disorder"). Gosset told Fisher he was still hopeful that K. P. would publish in *Biometrika* the new tables and their three papers.[5] Pearson had "warmed to the proposal." Fisher, for his part, "remained keen to retain the right of publishing elsewhere," namely, it turned out, in *Metron.*[6] And on July 17 Fisher wrote to Gosset, "I enclose the two notes I mentioned, the first of which is an attempt to give some idea of the multitude of uses to which your table may be put. . . . I have told Oliver & Boyd to send you two proofs [of *Statistical Methods*] as they become available. Many thanks for your offer. Yours sincerely, R. A. Fisher."[7] Pearson and *Biometrika,* and most of all Gosset, were being quietly skipped over.

So Fisher offered the tables and papers to Gini, keeping Gosset out of the loop. Eventually he informed Gosset of his plan, and Gosset was agreeable as usual. Three weeks later Gosset wrote again: "Dear Fisher, I am sending back your note and my new version which I hope is properly annotated. . . . As to the method of presenting the article [in *Metron*] whether under separate names or joint and the title (Somerfield rather boggles at the title I have put on mine!) I leave it entirely to you to do as you prefer and if necessary to put in liaison Material to putty up the joint."[8] As Florence David remarked, not a jealous bone in his body.

Fisher, thus empowered, did two things: first, he published three papers, individually, *not* jointly, in the sandwich order Fisher-Student-Fisher; and, second, without doing the "putty up" or making the matter of authorship of the tables clear, to repeat, he published the tables of *t* in his own article, the "Expansion" paper (1925c). In Gini's journal Fisher arranged for the collaboration a Gini coefficient, so to speak, of extreme inequality between him and Gosset.

Fisher published Gosset's tables under his own interpretation a third time, in 1938, with his former colleague and successor in the Department of Statistics at Rothamsted, Frank Yates, in *Statistical Tables for Biological, Agricultural, and Medical Research* (1938). In this widely distributed, much reprinted, and hugely cited book, Fisher and the as yet unreconstituted Yates failed to thank "Student" and *Biometrika*—let alone William Sealy Gosset—for permission to reprint the "table of *t*" (v–viii, 46). It is interesting to note that "R. A. F. and F. Y." signed the preface of the book's first edition in "August 1938" (v), less than a year after Fisher's old friend, the inventor and calculator of the *t*-distribution, had prematurely died. At the nearest opportunity, in other words—namely, at death—Fisher erased Gosset's name from Gosset's *t*-test and tables.

Reprinters of the *t*-distribution still thank "the Literary Executor of the late Sir Ronald A. Fisher, F.R.S., [and] to Dr. Frank Yates, F.R.S." (De-Groot 1975, v) or "Fisher & Yates . . . by permission of the authors and publishers" (693) or "the late Professor Sir Ronald A. Fisher, Cambridge, and Dr. Frank Yates, Rothamsted" (Hogg and Craig 1965, vi) and his and their publisher, "the Messrs. Oliver and Boyd, Ltd., Edinburgh, for permission to include [in *Introduction to Mathematical Statistics*] Table IV [i.e., "the *t*-distribution"] (371) or "Table III of Fisher and Yates . . . published by Oliver and Boyd, Edinburgh, and by permission of the authors and publishers" (Press 2005, ix, 533).

The late Frederick Mosteller called Samuel S. Wilks the "Statesman of Statistics" to honor the international leadership and public virtue of his gifted friend (1964). In the thirteen printings of Wilks's *Elementary Statistical Analysis,* "which," said Wilks, "at Princeton, is the introductory course for all fields of statistical application . . . designed for those who intend to go into the biological and social sciences . . . *Statistical Methods for*

Research Workers provided the *t* tables "by permission of the author, R. A. Fisher, and the publishers, Oliver and Boyd, Edinburgh" (1948, v, 208).

Leonard H. C. Tippett was a British statistician of note (he is perhaps best known for his once very useful "tables of random sampling numbers"). Tippett, who had known Gosset personally and had published an important article in 1928 with Fisher, saw some success with his own introductory textbook, *The Methods of Statistics*. Of the *t* tables printed in it Tippett writes, "These tables are taken by consent from *Statistical Methods for Research Workers*, by Professor Ronald A. Fisher, published by Oliver & Boyd, Edinburgh, and attention is drawn," says Tippett, "to the larger collection in *Statistical Tables*, by Professor R. A. Fisher and F. Yates, published by Oliver & Boyd, Edinburgh" (1952, 387). One almost detects in these famous textbooks by famous statisticians an eagerness to mention Edinburgh and Fisher. In fact DeGroot's *Probability and Statistics* and Hogg and Craig's *Introduction,* despite their undoubted merits as textbooks of mathematical statistics, each suggest, as do Tippett and Wilks and Press and many others, *twice*—in a preface and underneath the *t* table itself—that *t* originates with the Edinburgh-Fisher complex instead of with the *Biometrika*-Gosset complex or with the *Metron*-Gosset complex or with—truest of all—the Guinness-Gosset complex, as it did. Gosset has been scooped.

In 1937 Yule and Kendall correctly credited the tables to "Student" in their classic *Introduction to the Theory of Statistics* (538–39). More than Tippett and Wilks and the others, Yule and Kendall had lived near the vortex. Udny Yule himself—first-rate statistician, Hertz protégé, Karl Pearson hire, and K. P.'s first-ever student of statistics—had taught at University College, and was one of Gosset's many correspondents. But "Gosset" was, beyond Yule and a few other men and women of Gower Street, not a name that many other souls had ever heard. Fisher appears to have liked it that way. Inquiring letters to Fisher began to arrive in 1927 from overseas, with urgent questions about "his"—that is, Fisher's—new tables of *t*.[9]

Even the sophisticates Deming, Hotelling, and Jack Dunlap (a noted psychometrician at Stanford University) were unaware of the details of authorship. In *Some Theory of Sampling* (1950), the eminent Deming called Gosset's test, "Fisher's *t*" (1950, 541, 587). Deming himself wrote to Fisher, not Gosset, to clarify his own understanding of "Student's" *t*.[10] Sam Wilks did, too.[11] Hotelling, for his part, didn't once mention Gosset's name in a dozen printed reviews of *Statistical Methods* and *The Design of Experiments,* 1927 to 1939, in the *Journal of the American Statistical Association*.

Likewise in their own scientific articles Hotelling, Deming, and other Americans credited only Fisher and, following their example, so have many thousands of imitators worldwide.[12]

Fisher did not ever, it appears, acknowledge the mass confusion he had caused.[13] It would have anyway been uncharacteristic for him to do so. He would have had to admit straightforward scientific fraud.

Give Back to Gosset his Interpretation of Student's *t*

In short, in 1908 Gosset had published anonymously his two revolutionary papers, one on "The Probable Error of a Mean," the other on "The Probable Error of a Correlation Coefficient," inventing in them, among other things, his tables and test of z. In 1912 Fisher, while still an undergraduate at Cambridge, made a tiny correction to Gosset's z, and in 1922 they agreed to rename the corrected tables and test "Student's" t. Student copyrighted his own tables on three occasions, in 1908, 1914, and 1917. Fisher then copyrighted t in his own name and distorted it, in 1925 and again in 1938, turning it into a mechanical instrument for generating qualitative statements of existence. He suppressed Gosset's central statements about substantive, quantitative significance, "pecuniary value," and "real" error. Between 1925 and 1958 *Statistical Methods for Research Workers* was published in thirteen editions and many different languages, each successive edition in the English versions we have examined giving less and less credit to Gosset. The perhaps even more influential *Design of Experiments*, also widely translated and frequently reissued, was first published in 1935. In it Gosset's seminal insights on significance testing and the design of experiments are nowhere to be found. In an article of 1896, "Statistics of Unprogressive Communities," Edgeworth wondered, "How long does a wasp take in loading herself with sweets?"[14] We reply, about ten years.

Some have pointed out that in 1956 Fisher claimed that no scientist uses "a fixed level of significance" to "decide" a result (1956, 42)—and take this one sentence among his many hundreds of contrary statements to signify a change in his thinking.[15] But Fisher, the bulk of evidence suggests, was not earnest with it. He was playing a game. Fisher was in the 1950s understandably fearful of losing influence to the progressive Neyman and his "phantasy of circles" and to Egon, Wald, Deming, Shewhart, and Savage—the statistical sophisticates and champions of Bayes's decision theory.[16] In two letters of August 13 and August 16, 1955, to the British statistician G. S. James, Fisher expressed surprise that James or indeed any person would give a Neyman-Pearson interpretation of "the significance under which a composite hypothesis can be rejected." "I had thought," wrote Fisher, "that few other than Egon Pearson accepted this view. . . . It would therefore be useless to argue the matter" (quoted in Bennett 1990, 148–49). Besides, he told James, "I am averse to controversy in print or in letters" (149). Yet in 1962, the year of Fisher's death, he sent on March 13 a letter to another James. "I am not so sure that the tide of real feeling in the U.S. is so backward as it was 15 years ago," wrote Fisher, "when so many leading posts fell to Neymanians." He told James, "I think a good many [of the statisticians in the United States] are *dissatisfied* with *mere 'decisions,'* which essentially evade the problem of specifying the nature of uncertainty, or the true grounds for belief."[17] In a 1950 Eddington Memorial Address at Cambridge University, Fisher hammered away at his main point: "We attempt, so far as our powers allow, to understand the world, by reasoning, by experimentation, and again by reasoning. In this process moral or emotional grounds for preferring one conclusion to another are completely out of place" (1950, 15, quoted in Polanyi 1959, 153, n. 1). So he bent his principles, but did not break them.

THE HISTORICAL SIGNIFICANCE OF HOTELLING'S *t*'S

From the late 1920s to the mid-1960s, Harold Hotelling taught scores of students at Stanford University, Columbia University, and the University of North Carolina in economics, journalism, psychology, political science, and mathematical statistics. His own BA degree was, surprisingly, in journalism, which he kept a hand in. Oswald Veblen, Thorstein's brilliant mathematical nephew, supervised Hotelling's own PhD, at Princeton, in mathematics. But when Hotelling was hired for his first academic job, at the Stanford University Food Research Institute, he was expected to conduct statistical analyses. He didn't know how to. The director of the institute suggested he look at Fisher's *Statistical Methods*, just published.

By 1927 Hotelling and Fisher were in correspondence. And in 1929 Hotelling spent a semester sabbatical at the Rothamsted Agricultural Experimental Station, working side by side with the great man. Hotelling's statistical education became thoroughly Fisherian. For the next decade and beyond, the *Journal of the American Statistical Association* would be filled with appreciative reviews by α-Hotelling of the first seven editions of Fisher's *Statistical Methods for Research Workers* (Hotelling 1927 to

1939). By 1931 Hotelling had himself invented the Generalized T^2 ratio—a multivariate t-test—to the applause of Fisher (Hotelling 1931).

α-Hotelling became a vice president of the American Statistical Association and a leader of the movement to mathematize economics and political science. He made major contributions to economic and political theory as well as to statistics. Although he was not the most prolific of scientists, most of what he wrote was seminal. He published on the mathematics of demand theory, on welfare economics, on industrial organization, and on voting in two-party systems.[18]

But how to operationalize those ideas? In Hotelling's view, Fisher-significance was the way. Of the first edition of *Statistical Methods,* Hotelling writes: "Of particular interest are the methods of evaluating the significance of correlation coefficients drawn from small samples, the tests of significance of differences, and the method ([Fisher] 1927, 125) of fitting a polynomial to a series of observations by adding terms one at a time until the fit is sufficiently good. All these are due, I think, to Mr. Fisher's own researches [on the contrary, many of these, and certainly the main idea—came from Gosset]. Some of the tables," he continued, "particularly V(A) which gives the values that a correlation coefficient must attain in order to reach certain levels of significance, are indispensable for the worker with moderate-sized samples" (412); he added that the author, Ronald Fisher, must not be confused with the American economist, statistician, and eugenicist, *Irving* Fisher, or the Danish American actuary and statistician, *Arne* Fisher).

Hotelling's enthusiasm for the "brilliant contributions" (1927, 412) of Fisher impressed his immediate American contemporaries—Milton Friedman and W. Allen Wallis, for example, and their student Jimmie Savage.

> My statistical mentors, Milton Friedman and W. Allen Wallis, held that Fisher's *Statistical Methods for Research Workers* (1925) was the serious man's introduction to statistics. They shared that idea with their own admired teacher, Harold Hotelling. They and some others, though of course not all, gave the same advice: "To become a statistician, practice statistics and mull Fisher over with patience, respect, and skepticism." (Savage 1971a, 441–42)

While at Columbia University Hotelling and his wife hosted these and other talented students at regular "Hotelling *teas*," in obvious tribute to Fisher "significance" and Hotelling's own T^2, the students were made to understand.[19] Friedman and Wallis and Savage would each change their mind about Fisher. Like many of Hotelling's other students—Wald, Arrow, Kruskal, and Wolfowitz—Friedman and Wallis in time adopted a size-matters/how-much approach to inductive behavior. β-Friedman went even further, subscribing to Savage's notion of "personal probability" and Bayesian loss functions, both of which Fisher detested.[20]

Others saw Hotelling's devotion to Fisher in a rather different light. In May 1931 one of the other statistical Fishers, Arne, was living in New York City and received word that Ronald, whom he regarded as a friendly acquaintance, was "to give a series of lectures in Iowa City." He wrote to Ronald, hoping to get together with him in New York when his ship docked. In the letter Arne pointed to Hotelling as the chief example of Fisher's many "adoring disciples, who, in their effort to expand their own egos at the expense of their environment, are apt to overdo a good thing."[21]

> Take our friend Hotelling, for instance [Arne wrote]. According to him the adherents of the classical theory of probability must have sunk into a state of ease, almost approaching apathetic luxury, so that the time has become ripe for the Messiah, and lo, according to Hotelling, you are the savior to lead the statisticians out of the wilderness. . . . Why is it that your disciples are always so serious? Whenever I see them ascend the rostrum to expound—*ex cathedra* as it were—their pet dogmas in a solemn and gravelike manner, I cannot help thinking that they take themselves altogether too seriously and that humor could be used far more advantageously for administering a wholesome dose of truth. (A. Fisher 1931, 310–12)

One can only imagine the fury that such a letter would stir in the Wasp. He shot back to Arne: "On the whole . . . I would rather give my beard a Messianic cut, if this would encourage Shewhart and Hotelling to go ahead and do their job in a workmanlike manner."[22]

Each field had its messianic Fisher enthusiasts. Hotelling's more independent students were not at first prepared to challenge them.[23] In Europe and its empires the Fisher flag was first flown by Frank Yates and Udny Yule in biology, medicine, and social policy; by Arthur Bowley in economics; by J. Rasmusson in Sweden in genetics; by J. B. Hutchinson, at the Empire Cotton Growing Association, in Central India; and especially by P. C. Mahalanobis, who turned India Fisherian in agriculture, meteorology, physics, and anthropology, through the journal *Sankyhā*.

Fisher was conscious of his quest for empire. In October 1951 he wrote to the experimental psychologist W. E. Hick, "I am a little sorry that you have been worrying yourself at all with that unnecessarily portentous approach to tests of significance represented by the Neyman and Pearson critical regions, etc. [i.e., *Gosset*-significance]. In fact," said Fisher, "I and my pupils throughout the world would never think of using them."[24]

By the early 1950s a number of sophisticates were complaining. We have noted that in America many of Hotelling's students, and even some of his student's students, such as Savage, became apostates. In England the other inhabitants of Gower Street became even more vocal. After years of close collaboration with Fisher, Frank Yates of Rothamsted wrote in *Journal of the American Statistical Association* that Fisher's *Statistical Methods for Research Workers* "has caused scientific research workers to pay undue attention to the results of the tests of significance they perform on their data . . . and too little to the estimates of the magnitude of the effects they are estimating" (1951, 32). Fisher died in 1962. A year later Yates and the geneticist Kenneth Mather would let it rip in an unusually toughly argued biographical memoir of Fisher.

> It was the age of correlation and curve fitting. . . . It was also the age of coefficients of all kinds. In attempts to assess the degree of association in 2 × 2 contingency tables, for example, such measures as the coefficient of association . . . were proposed. The way in which these coefficients were used *revealed considerable confusion* between the problem of estimating the degree of association, and that of testing the significance of the existence of an association. *This confusion* permeated the whole of the statistical writing and thinking of the Pearsonian [and therefore, they mean, the Fisherian] school. (Yates and Mather 1963, 98; italics supplied)

"The whole" of it, they said. But Yates and Mather were just getting started. On pages 105–6 they returned to the topic of "significance" versus nonrandom "bias."

> At the time it was written (1925) [Fisher's] *Statistical methods for research workers* was a *tour de force*. Its main weakness . . . is insufficient consideration of the problems and pitfalls of estimation—bias in a regression coefficient due to error in the independent variate, for example, is not discussed—with the result that practical workers using Fisherian methods have often tended to place excessive emphasis on tests of significance, without asking whether the estimates they are testing are the appropriate ones.

And again, on page 113, speaking of "excessive emphasis" on "tests of significance" as against what should be "magnitudes of effects," in Fisher's *Design of Experiments:*

The main weakness of the book is that, as in *Statistical methods for research workers,* there is excessive emphasis on tests of significance:

"Every experiment [Fisher wrote] may be said to exist only in order to give the facts a chance of disproving the null hypothesis."

Considering the many experiments which are made *to estimate the magnitude of effects known to exist,* e.g. varietal differences, responses to fertilizers, this is surely a remarkable statement. (Italics supplied)

But the rebellions came too late. Fisher's 5 percent philosophy of inference had spread like the vine kudzu in the American South—it originated abroad; a salesman promised great prosperity from it in a variety of commercial, medical, and agricultural endeavors; once planted it grew and grew; and it suffocated everything in its path. Like kudzu, most people reckon that there's not much to be done about it.

Fisher declared in the tenth edition of his *Statistical Methods for Research Workers,* "To-day exact tests of significance [of the null hypothesis] need no apology" (1946, ix). That was an understatement. By then, Karl Pearson's 1 to 18 percent philosophy of science had evolved and hardened into Ronald A. Fisher's 1 to 5 percent ideology. There it has remained.

23

Eighty Years of Trained Incapacity:
How Such a Thing Could Happen

He knew there was something called "standard error." He remembered a professor of geology saying that the green-sand formation was so called because it was neither green nor sand; in the same way he thought "standard error" was so called because it was neither standard nor error.

SIR DANIEL HALL SPEAKING FOR E. S. BEAVEN
IN A 1936 DISCUSSION OF THE ROYAL STATISTICAL
SOCIETY CONCERNING GOSSET'S "CO-OPERATION IN
LARGE-SCALE EXPERIMENTS" (GOSSET 1936, 134)

Thorstein Veblen—Oswald Veblen's uncle, we noted, and so, astonishingly, the intellectual granduncle of Harold Hotelling—spoke in a famous phrase of the "trained incapacity" of a businessman to attend to anything but pecuniary profit. In fact Veblen used the phrase only once, on page 347 of *The Instinct of Workmanship* (1915), and on that page and page 193 in the alternative form, "trained inability." Neither phrase appeared anywhere else in his books, though the idea is fully consistent with Veblen's notion that sociological factors often overwhelm economic or especially engineering rationality. He used the idea of trained incapacity, if not the phrase, for example, in his other book of the same year, *Imperial Germany and the Industrial Revolution,* arguing (incorrectly, as it happens) that because of it the British railway freight cars were irrationally tiny.

The modest fame of the phrase (204 hits for ⟨Veblen "trained incapacity"⟩ on Google) appears to come from the prominence that the American literary critic Kenneth Burke (1897–1993) gave to it in his own book, *Permanence and Change* (1935), which was then picked up by the American sociologist Robert K. Merton. By the mysteries of Google there are in fact slightly more hits, 244, for ⟨Burke "Permanence and Change" Veblen "trained incapacity"⟩ than for the simpler ⟨Veblen "trained incapacity"⟩. Burke admitted in a letter in 1946 that he couldn't relocate the phrase in his Veblen collection after finishing his own book (1946). No wonder: as we said, it appeared only once in Veblen.

"Trained incapacity" was given by Burke a wider application than Veblen gave it. In Burke and others it means the narrowness imposed by a particular training. Our theme here has been that null-hypothesis significance testing is just such a trained incapacity—the incapacity, even, astonishingly, in the statistical works of the great Harold Hotelling, to see what is scientifically important because its victims have been trained to attend to what is not. "A way of seeing," writes Burke, "is also a way of not seeing—a focus upon object A involves a neglect of object B" (1935, 70). "Object B" in our case is "oomph."

It is our experience that the more training a person has undergone in Fisherian methods the less easy is it for her to grasp our very elementary point. Some of the direct students of Fisher, such as Hotelling, were nearly deaf to the point, seeing in Fisher only, as Fisher himself once put it, the "messianic cut." People who are highly trained in conventional econometrics—one or more generations descended from Hotelling and Haavelmo and Tinbergen and Klein—have an especially difficult time. Most of them have no idea what we are talking about, though they are quite sure they do not approve.

By contrast, undergraduates who have never had a statistics course, science and engineering professionals we work with or meet in our travels, businesspeople, musicians, activists, various colleagues in nonstatistical fields, and others who have the misfortune to ask what we are writing about these days, such as journalists for the *Economist, Strategy + Business* (Gabor 2006), the *Chronicle of Higher Education* and the like—as soon as they are able to grasp that we are *not* attacking statistics as such, and that we are not among the invincibly innumerate humanists unable to understand the Kolmogorov axioms—these have no difficulty understanding our point and immediately begin wondering what the controversy is about.[1] Physicists, for instance, get the point immediately and are not surprised to hear that the physics-trained Shewhart, Deming, Zellner, and Horowitz have always gotten it, too.

It is an interesting comment on the history of science that Fisher as a mathematical geneticist and statistician who is often credited with crafting a neo-Darwinian synthesis was himself an essentialist and creationist in his theory of scientific inference.[2] The sociological question is how such an error as the Rule of Two can persist. Or, rather, that is the *economic* question because sociologists have less trouble than economists do—another trained incapacity—in supposing that people can persist in gross ignorance year after year. Economists are likely to wonder why some smart person doesn't pick up the large-denomination bill lying on the ground and start a new intellectual firm, reaping the profits. If null-hypothesis significance testing is as idiotic as we and its other critics have so long believed, how on earth has it survived?

One possibility, which we have explored the most here, is the Great Scientist Thesis. Great scientists and their disciples have a great impact, despite defects. Fisher was undoubtedly a great scientist by this standard. He certainly had defects. One scale for measuring the bad and good effects of the great is our own field of economics. Economics since the 1940s has been dominated by Paul Samuelson. The Prudence-Only theory of human behavior that he and his brother-in-law Kenneth Arrow advocated has provided many interesting insights. But in the end it is as science a trifle strange. Perhaps more than a trifle. A world of economists without Samuelson and Arrow, splendid as both these scientists are, would have developed in a different and perhaps more reasonable—that is, in a way that invites to the center of the science more reasoned deliberation over human values—and especially a more intellectually pluralistic way. It might have nourished, for example, more DuBoises and Veblens, Douglases and Gilmans, Hayeks and Myrdals, Coases and Schellings—this despite the undoubted intellectual reasonableness of both Samuelson and Arrow as men and scholars (Arrow, we have noted, has long criticized null-hypothesis significance testing, and has, like Amartya Sen, maintained a second career in political philosophy), very unlike R. A. Fisher. It's the devout disciples in statistics and economics, as in other sciences, two or more generations descended from the master, one worries about.

Imagine, then, statistics without Fisher—suppose, for example, that his poor eyesight had not kept him out of the Great War he so fervently wished to attend, and suppose therefore, like so many of his generation of middle-class young Englishmen, he had died honorably as a promoted captain at the Somme. Statistics would have been very different. In particular, as we have suggested, the influence of Gosset would have shown through. With no Fisher to thwart them on Gower Street, Egon Pearson and Jerzy Neyman might have won the war over daily practice, not merely the battle for the theoretical high ground. Jeffreys' robust science could have reached the masses. Given the personal character and philosophical commitment to methodological pluralism of Gosset, Jeffreys, Egon, and Neyman, sans Fisher the "war" metaphor, applied to statistical practice, would probably have sounded silly. But to the bellicose Fisher not at all.

As we have also suggested, though, some strong divisions were in fact brought into play against the Fisherian troops early and late. Deming and Gosset, for example, took a substantive approach to significance into the business world and made it work. Wald, Wallis, Savage, Lindley, Good, Lehmann, Kruskal, and many others in advanced theoretical statistics advocated substance over Fisherian significance, and a minority of their own students persisted in the higher training. But the mass of research workers stayed Fisherian, lost in a blindingly sizeless stare. With such enemies, and so many decades now passed after Fisher's death, you would think that the effect of the great scientist would begin to fade.

But scientific hubris is a stubborn cat. Another scientific error of the early twentieth century, Freudianism, for example, is by now quite dead as scientific psychology and has been for decades. Only very elderly psychoanalysts still believe that everyone is a repressed homosexual or forever struggling with Daddy. Yet Freudianism is vividly alive still in the work of literary theorists, especially in France and the United States. Something besides the personal influence of Freud himself and his most obedient disciples must be explaining such persistence. Without Freud, no Freudianism. Without Fisher, no Fisherianism. But given Freud and Fisher, something other than their charisma must be explaining the persistence of their erroneous ideas.

Of course, one can insist stoutly that their ideas are *not* erroneous. The Efficiency Thesis would say that null-hypothesis significance testing is *necessary* for drug testing and economic forecasting and other daily-delivery tasks. We hope by now that you do not believe this. To refer one last time to what Deming said in 1938, statistical significance in fact provides no rational plan of action. True, the price of significance tests delivered to your doorstep every morning like milk is temptingly low. One can therefore take null-hypothesis significance testing as a sort of astrology, giving "decisions" mechanically, justified within the system of astrology itself. (A substantial fraction of French business firms, by the way, still consult astrologers.) Statistical Package for Social Sciences and other such companies sell data mining to business—one SPSS product is called "Clementine," after the American folk song about "a miner, forty-niner / And his daughter, Clementine," honest anyway in the naming.

Data mining, we affirm, is not in itself scandalous. Mere untheorized cataloging of the stars by location and magnitude allowed the astronomers of 1572 and 1604 to be astonished by novae and to infer that the heavens were not in fact immutable. Close attention to the bundles of goods its customers buy—a kind of data mining—has made Wal-Mart rich. But data mining guided by considerations of 5 percent significance, as in fact it is in Clementine and other statistical software, is worse than data mining guided by astronomical or retail-giant significance. With significance testing in the age of digital computing, prices are always falling but so is scientific knowledge. Fisherianism is a *bad* input, straightforwardly misleading advice, erroneous astrology. Misleading advice is not made into good advice merely by its mechanical and pecuniary cheapness.

Another explanation along similarly pro-Fisherian lines is that Precision *Is* Best. Again, we hope by now you are as impatient with such an argument as we are. Testimators rest content with a nominal level of statistical significance, ignoring the real significance—the rise or fall in the price of the ostensible object of inquiry. Suffering from precision illusion, they ignore real error. The economic approach to the logic of uncertainty is by this camp rejected "owing," it may be as de Finetti put it, "to aristocratic or puritanic taboos."

Life and human scientists have been for a long time—at least since the time of Laplace—intrigued by and even obsessed with the precision and control they believe to exist in the so-called exact sciences of physics and chemistry. Sir Ronald Fisher certainly was. Even today the philosophers of economics and medicine and the others yearn for a demarcation criterion segregating the exact and the inexact.[3]

But statistical significance, we hope by now we have persuaded you, is not the instrument of exactitude it pretends to be. Cell biologists and molecular geneticists and medieval economic historians have improved the exactitude of their sciences, that is true. Great strides have been made since the 1950s. Yet statistical significance has not done it.

A more satisfactory explanation is Path Dependence. Once Fisher had got the ball rolling, it continued hurtling down, gathering speed and influence. Almost all the textbooks, as we have seen, recommend a simpleton's version of Fisher—or, as Gigerenzer et al. have shown, at best an incoherent hybrid of the Fisher-Bayes-Neyman-Pearson setup followed by full indulgence of Fisher's testimation.

Path Dependence, though, is merely a catchall for factors that keep people from attending to social costs, or keep them from taking advantage of the opportunity even for private profit lying on the ground before them. What makes the path "depend?" As with the QWERTY layout of the keyboard we are typing with, considering the great ease of changing keyboards in these days of computers, and the potential gains to typing-intensive companies in doing so, there have been profitable alternatives to Fisherian significance all along. Some manner of *other* sociological factor must be intervening.

One of the sociological factors, as we have said, is High Modernism. Fisher had the good fortune to be born just as the prestige of mechanical methods in all fields, from mathematics to automobile manufacturing, was coming to a climax. Methods with a capital *M*, formulas for architecture or economic planning, were believed during the early twentieth century with a highly nonmechanical passion. Medicine, biology, social work, and the social sciences were just developing their national associations, developing standards, when Fisher emerged. Now in the early twenty-first century some are less persuaded.

The other sociological factor, itself connected to Modernism, is what Robert Merton called the Bureaucratization of Knowledge. In his *Social Theory and Social Structure* (1949), quoting Burke on Veblen, he spoke of the increasing systematization of modern life in large organizations. "Adherence to the rules," Merton wrote, "originally conceived as a means, becomes transformed into an end-in-itself" (1949, 199). That seems about right: statistical significance, originally conceived as a means to substantive significance, became transformed by Fisher and then by bureaucracies of science into an end in itself. A *t*-tested certified fact will be "equally convincing to all rational minds, irrespective of any intentions they may have in utilizing knowledge inferred." Seek it. "Ignore all results which fail to reach this standard." The modern-day Gradgrinds intone, "Teach these boys and girls in economics, medicine, ecology, nothing but *statistical* significance. Plant nothing else, and root out everything else."

Merton speaks of the "transference of the sentiments from the *aims* of the organization onto the particular details of behavior required by the rules." And: "Formalism, even ritualism, ensures with an unchallenged insistence upon punctilious adherence to formalized procedures. This may be exaggerated to the point where primary concern with conformity to the rules interferes with the achievement of the purposes of the organization" (1949, 200; italics his). The ostensible purpose of a scientific organization is to produce knowledge. In the sizeless sciences, the formalism, even the ritualism, of the Rule of Two or Three and the transposed conditional interfere. The "rules in time become symbolic in cast," writes Merton, "rather than strictly utilitarian." Alas, yes.

If we were to assemble our socioeconomic observations into a single chain of thought its strongest link would be coupling Merton's "bureaucracy" with Hayek's "scientism." *Scientism* describes, "of course, an attitude which is decidedly unscientific in the true sense of the word, since it involves a mechanical and uncritical application of habits of thought to fields different from those in which they have been formed" (Hayek 1952, 15–16). "The scientistic as distinguished from the scientific view is not an unprejudiced but a very prejudiced approach which, before it has considered its subject, claims to know what is the most appropriate way of investigating it."

The trick is to unshackle the bureaucracy of scientism, to break its mechanical rules, change its prejudiced incentives, create new rituals, train capacity. No simple trick.

24

What to Do

Not where it comes from but what it leads to is to decide.
WILLIAM JAMES 1896, 98

EDUCATION

Our first suggestion is mild and educational. Scientists in the sizeless sciences need to start telling each other to seek substance. They need to stop believing that the translation of a problem into probability space relieves them of the need to consider oomph and loss functions. The ritualism of significance testing needs to be challenged at the level of paradigm at every seminar, in every referee report, in every classroom.[1] The physicist and econometrician Joel Horowitz claims to us that such challenges have become common in economics, partly because of our earlier complaints. We think Horowitz is mistaken. The Fisherian ritual goes on and on. Follow then Horowitz's own practice: What's your oomph? How do you know?

Listen to Gosset explaining in his last year of life his thoughts on "significance" to Egon Pearson, who was then the chief editor of *Biometrika*.

> [O]bviously the important thing in such is to have a low real error, not to have a "significant" result at a particular station. The latter seems to me to be nearly valueless in itself. . . . Experiments at a single station [i.e., tests of statistical significance on a single set of data] *are* almost valueless. . . . What you really want is a low real error. You want to be able to say not only "We have significant evidence that if farmers in general do this they will make money by it," but also "we have found it so in nineteen cases out of twenty and we are finding out why it doesn't work in the twentieth." To do that you have to be as sure as possible which is the 20th—your real error must be small. (Gosset to E. S. Pearson, 1937, in E. Pearson 1939, 244)

Gosset-speak is what we need. Undergraduates need to hear from the beginning that size matters—measured in units of money or justice or life or persuasiveness. They need to acquire the virtues necessary for performing repeated experiments on the same material. They need to hear that random error is one out of many dozens of errors and seldom the biggest. They need to learn that "the *real* error must be small." After all, reconciling differences of effect, finding the common ground, is the point of statistics. Professors should show students why they need to attend to substance, as does for example the epidemiologist Kenneth Rothman. The point about the insignificance of significance should not be shunted off to one obscure paragraph mentioning that large samples yield "significance" everywhere. As in the Freedman, Pisani, and Purvis text, or in the old text by Wallis and Roberts (1956), and in the careers of Gosset and W. Edwards Deming, the size-matters/how-much should be the substance of most of the paragraphs.

Controlling for the second kind of sampling error is necessary and important. But it is not most important. Most important is to minimize Error of the Third Kind, "the error of undue inattention," which is caused by trying to solve a scientific problem using statistical significance or insignificance only. In science, as against careerism or pure mathematics, it is better to be approximately correct and scientifically relevant than it is to be precisely correct but humanly irrelevant. Not even the fully specified power function, balancing the risk of errors from random sampling, provides a full solution to a scientific problem. In truth, as Kruskal never tired of remarking, statistical "significance" poses no scientific problem at all. With the aid of a personal computer and a grant such significance is easy to achieve.

Graduate students today are oversupplied with analytic proofs of asymptotic results. They are not being taught how to control for the third kind of error. Formalities are privileged in textbooks over substantive thinking about what a test can yield. We have met too many well trained young economists who say to us, "I didn't know I was supposed to look at the column of coefficients. I just look at the *p*-values." And "I thought Fisher's test told me about the likelihood of the hypothesis. Doesn't it?" No.

If Gosset could teach these points to the elder Beaven, an experimental farmer of barley wisely suspicious of the logic of Latin squares and the importance of degrees of freedom in tests of statistical significance— Beaven protested wittily to Gosset that he did not understand "magic squares" or "birds of freedom" (McMullen 1939, 208)—we can certainly teach them to our mathematically savvy students. Statistical scientists can teach substance without sacrificing the rigor they so passionately seek. Real rigor will *rise* with increased attention to substance.

It is rare to meet a statistical scientist who knows about the rhetorical history of Student's *t* and rare at the .001 level to meet one who knows the science of Student himself. For eighty years the Wasp has won the battle for historiographic attention. A few historians and philosophers of statistics—Hacking and Stigler and Giere and Mayo and Howie and Zabell, for example—have begun to question the omission of alternative approaches. Yet to matters of substantive significance the historians and philosophers have not quite twigged.

The exceptions, those who have critically examined the omission of substantive significance and who understand well the ongoing accumulation of social damage, are in fact applied statisticians and statistical theorists, not historians or philosophers. We mean, in particular, Savage (1954, 1971a), Kruskal (1968a, 1980), Zellner (1984, 1997), Gigerenzer et al. (1989), and Berger (2003). The Savage article of 1971, "On Rereading R. A. Fisher" (1971a) is unknown outside a small circle of statisticians. It should be reprinted and put into the hands of every graduate student of the statistical sciences. Gigerenzer et al., *The Empire of Chance* (1989), has been noticed favorably by historians of science. Its brief history of the birth of mathematical statistics in its Fisherian and then its Fisher-Neyman-Pearson hybrid form is excellent. But on the main point of substantive versus statistical significance, in emphasizing the distinction and why the distinction matters, the Gigerenzer book is—like the illuminating book of Donald MacKenzie, *Statistics in Britain* (1981)—not as explicit as it might be.[2] We were alarmed to find a scholar and statistician of the quality of Bradley Efron saying that "we could badly use a new Fisher to put our world in order" (1996, 112).[3] Another Wasp is precisely what we do not need.

STOPPING TESTIMATION

Our second suggestion is harsh and institutional. Somehow the institutional incentives that are leading the sizeless sciences away from oomph need to be changed. We know enough about the bureaucracy of scientism to have doubts that educational reform alone will suffice.

The mildest form of our harsh suggestions is to ask our honored colleagues in statistics and economics and medicine and the rest to stand up and be counted. Consider econometrics. We know, for example, that the Nobel laureate James J. Heckman of the Department of Economics at the University of Chicago agrees with us in substance *about* substance. His own practice shows it. We now ask him and similarly placed men and women to follow the recent example of Thomas Schelling ([2004], Nobel laureate, 2005) and say publicly what is obvious and elementary: statistical significance is not the same thing as importance, precision is not the same thing as oomph, but measures of oomph and importance are what we mainly need. The current practice of statistical journals, we hope you now agree, is mistaken. Testimation—the dysfunctional marriage of the sizeless stare and the fallacy of the transposed conditional—should cease. If six economic scientists of the stature of Heckman would join their fellows Arrow and Zellner and Schelling and Solow and Granger and Leamer and the rest in a forthright statement of the absurdity of confusing precision with oomph, econometrics might yet be saved.

Our proposition applies especially to the editors of the journals. If journal editors cannot answer us—we have been awaiting enlightenment from them these decades past—they should in all scientific seriousness admit that none of the "tests" they publish makes sense. That a journal editor would ask a leading economist and applied decision theorist such as the late Jack Hirshleifer to compute and publish Fisherian test statistics that he did not believe in is most strange. "I'm just baffled that anyone, anytime, anywhere, can think it scientific to report only Type I errors without reporting the magnitude of the effect," Hirshleifer wrote to us a year before his death (Hirshleifer 2004). "To my mind," he continued, "the Bayesian objection would emphasize the need to account also for Type II errors. Of course just how you account for them will depend upon the loss functions, and that's where 'size' comes in." Journal editors should be embarrassed. The editors might say, à la Rothman: "Dear Author: Our journal does not publish articles claiming statistical significance to be the same as scientific importance. Your article, in its current version, makes such a claim. So if you would like to publish here you have two courses of action . . ."

We will ask journal editors to sign a cold-water pledge, and will print the results of our survey—if, indeed, we can find an editor who will publish them. Journals with low standards in such matters, such as the *American Economic Review, New England Journal of Medicine, Annals of Internal Medicine, Lancet, British Medical Journal, Biometrika, Psychometrika, Administrative Science Quarterly, Journal of the American*

Medical Association, and *Journal of Clinical Psychiatry,* should be made known to the scientific community. Statistical significance does not an "impact factor" make. Against testimation we propose a "Statement on the proprieties of Substantive Significance" (SpSS), and will ask editors, administrators, and eminent scientists and people of industry concerned with such matters to sign.

(1) Sampling variance is sometimes interesting, but a low value of it is not the same thing as scientific importance. *Economic* significance is the chief scientific issue in economic science; *clinical* significance is the chief issue in medical and psychiatric and pharmacological science; *epidemiological* significance is the chief issue in infectious disease science; and *substantive* significance is the chief issue in any science, from agronomy to zoology. No amount of sampling significance can substitute for it.

(2) In any case, scientists should prefer Neyman's *confidence intervals,* Rothman's *p*-value *functions,* Zellner's *random prior odds,* Rossi's *real Type I error,* Leamer's *extreme bounds analysis,* and above all Gosset's *real error bars* (Student 1927) to the Fisher-circumscribed method of reporting sampling variance (Leamer 1982; Leamer and Leonard 1983; Zellner 2008). No uniform minimum level of Type I error should be specified or enforced by journals, governments, or professional associations.

(3) Scientists should prefer *power functions* and *operating characteristic functions* to vague talk about alternative hypotheses, unspecified. Freiman et al. (1978), Rossi (1990), and similar large-scale surveys of power against medium and large effect sizes should serve as minimum standards for small and moderate sample size investigations. Lack of power—say, less than 65 percent for medium-sized effects and 85 percent for large effects—should be highlighted. How the balance should be struck in any given case depends on the issues at stake.

(4) *Competing hypotheses should be tested against* explicit *economic or clinical or other substantively significant standards.* For example, in studies of treatments of breast cancer a range of the size and timing of expected net benefits should be stated and argued explicitly. In economics the approximate employment and earnings rates of workers following enactment of a welfare reform bill should be explicitly articulated. Is the Weibull distribution parameter of the cancer patient data substantively different from 1.0, suggesting greatly diminishing chances of patient survival? How greatly? What does one mean by the claim that welfare reform is "working"? In a labor supply regression does β = "about -0.05" on the public assistance variable

meet a defensible minimum standard of oomph? In what units? At what level of power? Science needs discernible Jeffreys' d's (minimum important effect sizes)—showing differences of oomph. It does not need unadorned yet "significant" t's.

(5) *Hypothesis* testing—examining probabilistic, experimental, and other warrants for believing one hypothesis more than the alternative hypotheses—should be sharply distinguished from *significance* testing, which in Fisher's procedure *assumes* a true null. It is an elementary point of logic that "If H, then O" is not the same as "If O, then H." Good statistical science requires genuine hypothesis testing. As Jeffreys observed, a p-value allows one to make at best a precise statement about a narrow event that has not occurred.

(6) Scientists should estimate, not testimate. Quantitative measures of oomph such as Jeffreys's d, Wald's "loss function," Savage's "admissibility," Wald and Savage's "minimax," Neyman-Pearson's "decision," and above all Gosset's "net pecuniary advantage" should be brought back to the center of statistical inquiry.

(7) Fit is not a good all-purpose measure of scientific validity, and should be deemphasized in favor of inquiry into other measures of error and importance.

Social indignation can be used for social gain, as Adam Smith long ago noted. Fisher's contributions to science are many and great. In genetics, agronomy, and mathematical statistics his name will endure. But Fisher Significance and its cousin, the fallacy of the transposed conditional, should go the way of bloodletting.

Paul Green, a colleague of Ziliak's at Roosevelt University, a political scientist with sophistication in the history of institutional change, sees a daunting parallel in the Old Left. Returning to Chicago after the last re-election of Tony Blair, Green remarked, "Some parties [such as Britain's Old Left, defeated by Blair] don't *want* control; they don't *want* to win. In Chicago, winners take over City Hall, and celebrate," said Green. "Losers provide explanations. The Old Left in Britain," as Green sees it, "would rather provide explanations." Significance testers are Old Left testimators—they resemble the political Old Left, or the Democratic poll watchers in Florida in the year 2000: they would rather provide explanations than regain control.

The textbooks are wrong. The teaching is wrong. The seminar you just attended is wrong. The most prestigious journal in your scientific field is wrong.

You are searching, we know, for ways to avoid being wrong. Science, as Jeffreys said, is mainly a series of approximations to discovering the sources of error. Science is a systematic way of reducing wrongs or can be. Perhaps you feel frustrated by the random epistemology of the mainstream but don't know what to do. Perhaps you've been sedated by significance and lulled into silence. Perhaps you sense that the power of a Rothamsted test against a plausible Dublin alternative is statistically speaking low but are dazzled by the one-sided rhetoric of statistical significance. Perhaps you feel oppressed by the instrumental variable one should dare not to wield. Perhaps you feel frazzled by what Morris Altman (2004) called the "social psychological rhetoric of fear," the deeply embedded path dependency, that keeps the abuse of significance in circulation. You want to come out of it. But perhaps you are cowed by the prestige of Fisherian dogma. Or, worse thought, perhaps you are cynically willing to be corrupted if it will keep a nice job.

Repent, we say. Embrace your inner Gosset. Appeal to the ancient argument of *sorites*. Sell all your goods and come with us. As the comedian Richard Pryor put it, "Who are you going to believe—*us* or *your own lying eyes?*"

A misleading rhetoric brought in the age of mechanical testing and with it a lot of damage. Good rhetoric can replace it, and take us into the age of science and humanity and, if chance is on our side, into another fine stout.

A Reader's Guide

Although our approach is mostly nontechnical, we assume the reader is broadly familiar with testing, estimation, and error statistics as used in the life and human sciences. Some readers may appreciate guidance along these lines. Two good, nontechnical introductions to the topics discussed in this book are David Moore and George McCabe's *Introduction to the Practice of Statistics* (1999) and J. Pratt, H. Raiffa, and R. Schlaifer's *Introduction to Statistical Decision Theory* (1995). The second edition of Kenneth Rothman's *Epidemiology* (2002) and his advanced *Modern Epidemiology* (1986) are especially relevant to research workers in allied fields such as medicine, psychiatry, pharmacology, and even some parts of economics. Psychologists and education researchers may find instruction, as have we, in Bruce Thompson's *Foundations in Behavioral Statistics: An Insight Based Approach* (2006).

The student of how size matters and what to do about it in the fields of economics and other human sciences would do well to begin with *Introductory Econometrics* (2000), by Jeffrey Wooldridge, which gets to the point. In the 1950s the great Polish economist Oskar Lange advocated "practical" or "economic" significance along similar lines in the second edition of his *Introduction to Econometrics* (1959). Although Lange's book is out of date, technically speaking, it is still a model of real world econometrics. We wish his views on economic planning had been as sensible.

At the level of foundations, Fisher is nothing like the last word. "Personal probability," in the tradition of Gosset and Leonard Savage, is an idea that has not received its due. Yet it is a particularly natural aid for making decisions in the fields of medicine and economics. Graduate students will profit to that end from Savage's *The Foundations of Statistics* (1954 [1972]), S. James Press's *Subjective and Objective Bayesian Statistics*

253

(2003), and two of Arnold Zellner's collections (which are sympathetic with, but not strictly devoted to, the personal approach): *Basic Issues in Econometrics* (1984) and *Bayesian Analysis in Econometrics and Statistics* (1997). Add to these books Tony Lancaster's hands-on *Introduction to Modern Bayesian Econometrics* (2004)—though treat the sections on "significance" with deep suspicion—Press's *Applied Multivariate Analysis* ([1972] 2005), and Leamer's *Specification Searches: Ad Hoc Inference with Non-experimental Data* (1978) and you're equipped to make persuasive econometric arguments. Students may want to see well-written examples of oomph*ful* science. Our book supplies, we think, a few examples. We can recommend M. E. Bowen and J. A. Mazzeo, eds., *Writing about Science,* a collection of nontechnical pieces written by famous scientist-essayists. In the essays by Richard Feynman, Victor Weisskopf, Lewis Thomas, and Howard Ensign Evans, oomph is the word.

The student in search of an elegant finish to her statistical inquiries could do well to consult "The Art of Labormetrics," by Daniel Hamermesh (1999). The old classic by Oskar Morgenstern, *On the Accuracy of Economic Observations* (1950 2nd ed., 1963), is bracing. Hamermesh and Morgenstern are to word and number what Edward Tufte is to visualization: they offer a corrective to the output and display of statistics presently determined by the default settings of your software system. The common sense of a brewer, of course, completes the job.

Aris Spanos has recently brought the Fisher-Neyman-Pearson-Jeffreys debates into econometric learning in his *Probability Theory and Statistical Inference* (1999). We salute the inclusion of such pluralism in the dismal science. Although the author appears to disagree with us on the logic of uncertainty and on some of the history, we find exemplary the open-minded character of his book. As the rhetoricians have put it since Gorgias of Leontini, there is not just one way.

Notes

A Significant Problem

1. Student 1908a, 1908b; E. Pearson 1939, 222–23; Ziliak 2008a, 2008b; cf. Stigler 1999, 143–44.

2. A helpful introduction to the conditions of "work and welfare" at Guinness, 1886–1914, is Dennison and MacDonagh 1998, chap. 8. At Guinness a junior brewer earned far more than the average London professor of science. Gosset stopped Karl Pearson from making a job offer to him for precisely that reason (E. Pearson 1990, 18).

3. Letter, W. S. Gosset to Karl Pearson, September 1, 1915, reprinted in E. Pearson 1990, 19.

4. Student 1905, reprinted in E. Pearson 1939, 215–16; Neyman 1938, 1957; Jeffreys 1939a; Savage 1954, 1971a; Leamer 1978; Press 2003; Fisher 1955, 1956; Popper 1959.

5. Laplace (1773), quoted in Scott 1953, 20. On the relation of welfare programs to economic growth see, for example, Atkinson 1999.

6. See instead Student 1908a, 1908b, 1927, 1931a, 1938, 1942; Gosset 1936; Pearson and Adyanthaya 1929, 276; Morgenstern 1950, chap. 2; Deming 1950, chap. 2 [the spirit of Deming's *Some Theory of Sampling* is, despite the title, *mostly* about how to control for some *twenty* different kinds of *nonsampling error*]; Rothman 1986, chaps. 7–16; Mayo 1999, 4–7; Mosteller 1968, 208–29; Elashoff and Elashoff 1968, 229–50; and Heckman 1979. "Sampling error" is a tiny subset of what Gosset and these others call "real" or "total" or "actual" error.

7. Letter, Harold Jeffreys to Ronald A. Fisher, February 24, 1934, reprinted in full in Bennett 1990, 150.

8. Letters 1–2, W. S. Gosset to E. S. Pearson, 1926, in Pearson Papers, Egon file, Green Box, Special Collections, University College London (hereafter UCL); Wald 1950; Savage 1954, 156–57, 163–83; Chernoff 1968, 131–35.

9. Reid 1982, 68–70.

10. Savage 1954, 6; italics supplied.

11. Ibid., 6.

12. See, for example, Tversky and Kahneman 1971; and Altman 2006.

13. Quine and Ullian 1970, 12–13.

14. Karl Pearson ca. 1905, 15, "Dice," Box 67, Pearson Papers, Karl file, UCL.

15. Savage 1954, chap. 10.

16. Student 1927, 1931a; Gosset 1936.

17. Steward and O'Donnell 2005.

18. "William Sealy Gosset," Guinness Archives, GDB/C004.06/0001.04; "Research Papers from Second Volume of the Brewery History—Brewers and Directors," Guinness Archives, GDB/C004.09/0004.14.

19. Letter, W. S. Gosset to E. S. Pearson, May 11, 1926, "Student's Original Letters," Letter 1, Green Box "G7," Pearson Papers, Egon file, UCL, reprinted in E. Pearson 1939, 243.

20. See, for example, Fisher 1914, 1915b.

21. Egon S. Pearson was the official editor of *Biometrika* from 1936 to 1966. Beginning in 1933, however, Egon did much of the editorial work because of Karl's cataracts and increasingly poor health.

22. Beaven, a self-trained man who began to farm and malt barley at the age of thirteen, was awarded in 1922 an honorary doctorate from Cambridge University "in recognition of his work on crop improvement" (Hugh Lancaster, "Edwin Sloper Beaven (1857–1941)," Museum of English Rural Life, DBEAV A3). The only comprehensive introduction to his work on malting and experimental agriculture is E. S. Beaven, *Barley: Fifty Years of Observation and Experiment* (1947).

23. Letter 6, W. S. Gosset to E. S. Pearson, April 19, 1937, Pearson Papers, Egon file, UCL (italics supplied; reprinted in E. Pearson 1939, 247). Gosset was preparing at the time what would become his last published paper, "Comparison between Balanced and Random Arrangements of Field Plots" (Student 1938). To see the entire Gosset quote, which grows increasingly radical in its antisignificance rhetoric, turn to the beginning of chapter 24.

24. Student 1905, 219.

CHAPTER 1

1. *K. P. Lectures, Volume I* [Gosset's classroom notebook], 13, 1906–7, Pearson Papers, Gosset file, UCL.

2. Berenson 2005, A14; italics supplied.

3. W. S. Gosset, "Report on the Application of the 'Law of Error' to the Work of the Brewery," Arthur Guinness Son and Co., November 3, 1904, reprinted in E. Pearson 1939, 213–14; italics supplied.

4. See, for example, Kruskal 1968a, 238–50; and DeGroot 1975, 496–97. Both Kruskal and DeGroot emphasize the need, therefore, to consider the power of the test, as Kruskal put it, "even if only crudely and approximately."

5. See, for example, "Eugenics in Britain," chap. 2 of Donald MacKenzie's *Statistics in Britain, 1865–1930* (1981); and Daniel J. Kevles's *In the Name of Eugenics* (1985). For the prehistory in economics, see David M. Levy's *How the Dismal Science Got Its Name: Classical Economics and the Ur-Text of Racial Politics* (2004); and Sandra J. Peart and David M. Levy's *The "Vanity of the Philosopher": From Equality to Hierarchy in Post-Classical Economics* (2005).

CHAPTER 2

1. Deming's size-matters/how-much approach to the philosophy of inference is discussed in chapter 10.

CHAPTER 3

1. Savage 1954, chaps. 8–10, 15–16, is the place to start.

2. The oomph-conscious pharmacologist will succeed, that is, if he is not forced by a government regulatory agency to reject all experiments with $p \geq .05$. This is part of what we mean in chapter 23 when we say that the *bureaucracy of scientism* must be reformed.

3. This is Galton's "Statistical Inquiries into the Efficacy of Prayer" (1872). The article has been read by religious conservatives as an attack on prayer. We wonder if such critics have read the article, especially its final two paragraphs on the personal and spiritual meanings of prayer.

4. Savage 1954, 252–56, 167–77.

5. Howie 2002, 92–93; Zellner 2004a.

6. Wrinch and Jeffreys 1921; Jeffreys 1919.

7. The strongest statements in a strictly statistical context are, we think, in Neyman and Pearson 1928, 1933.

8. Leamer 2004, 556; italics in original.

9. Cicero *Academica* 93, trans. II. Rackham, 1951, 586, in Black 1970, 1–2. Also see Lanham 1991, 143.

CHAPTER 5

1. Jeffreys 1967, 384, quoted in Zellner 1984, 277n. Cf. McCloskey 1985b, chap. 7.

2. *American Economic Review,* June 1981, 298, n. 11; 305, n. 27.

3. Shiller 1987; Solow 1988, 183–204.

CHAPTER 6

1. For further ominous "KP" results, see chapters 18 and 19 in this volume.

CHAPTER 7

1. Zellner 2005b; italics in original.

2. E. Pearson 1990, 11–13.

3. Quoted in Morgan 1990, 254, n. 15.

CHAPTER 8

1. Genberg 1976; McCloskey and Zecher 1976.

2. Kravis and Lipsey 1978, 204–5, 235, 242.

Chapter 10

1. Griliches and Intriligator 1983–86, 1:325.

2. See Tyler 1931; Shulman 1970; Carver 1978; and scores more, from Boring to Meehl to Cohen to Schmidt to Thompson, already mentioned.

3. Edgeworth 1885, 1907; Wald 1939; De Finetti 1971; Savage 1971a, 1971b; Cain and Watts 1970, 229, 231; Wooldridge 2000, 131–34; Keuzenkamp and Magnus 1995; McCloskey and Ziliak 1996, 99 and numerous other references on 112–14; McCloskey's citations in her "works cited" sections in McCloskey 1985a, 1985b, 1992, 1995, 1996, 1998); Darnell's comprehensive review of 1997; Hamermesh 1999; Colander 2000; Keuzenkamp 2000, 266; Berg 2004; Schelling 2004; and Ziliak 2004, 2005, 2008a.

4. Hotelling 1927 to 1939.

5. See Morgan 1990, 242, n. 10, which claims, on the contrary, that Haavelmo "adopted" the Neyman-Pearson procedures. See also the Haavelmo reference in Savage 1971b, 111–12.

6. Vining 1949, quoting from Yule in the *Journal of the Royal Statistical Association* n.s. 105, Pt. II, 83ff.

7. Klein (1943), in Klein 1985, 35.

Chapter 11

1. Atlas 2000, 263–66.

2. *Publication Manual of the American Psychological Association* (hereafter *APA Manual*) 1952, 414, quoted in Fidler et al. 2004, 619.

3. *APA Manual* 1994, 18; Thompson 2004, 608; Fidler et al. 2004a, 619.

4. *APA Manual* 2001, 22; Thompson 2004, 609.

5. Hill and Thompson 2004, cited in Fidler et al. 2004a, 619.

6. Freedman, Pisani, and Purves 1978, A23.

7. Fisher 1955 (cf. Student 1927; and Gosset 1936); Ziliak 2008a.

8. Hubbard and Ryan 2000, fig. 1 in Fidler et al., 2004a, 618.

9. Thompson 2002b, 2004; Schmidt 1996.

10. Wilkinson and APA Task Force 1999, 599.

11. Fidler et al. 2004b, 119.

12. Vacha-Haase et al., 2000, 419, italics deleted; quoted in Fidler et al., 120.

Chapter 12

1. Letter, W. S. Gosset to E. S. Pearson, May 11, 1926, Letter 1, Pearson Papers, Egon file, Green Box, UCL; emphasis in original; partially reprinted in E. Pearson 1939, 243.

2. Neyman and Pearson 1933, 296.

3. Cf. Bakan 1966; and Tversky and Kahneman 1971.

Chapter 13

1. Boring 1929; Gigerenzer et al. 1989, 205–14; Howie 2002, 193–95.

2. Howie 2002, 194.

3. Adkins 1968, 22.

4. Gigerenzer et al. 1989, 205–14; Howie 2002, 194.

5. Porter 1995, 3–8; Cohen 2005.

6. Lowe and Reid 1999, 15–17.

7. Guilford 1942, 217.

8. Gigerenzer 2004, 587–88.

9. Fisher 1955, 70; Savage 1971a, 464.

10. Reid 1978, 182–83, 264–66.

11. Arrow, Blackwell, and Girshick 1949. Blackwell and Girshick went on to publish a well-regarded book on game theory, *The Theory of Games and Statistical Decisions* (1954). Girshick and Savage collaborated, too. See, for example, their "Bayes and Minimax Estimates for Quadratic Loss Functions" (1951).

12. Blackwell, n.d. (www.amstat.org/statisticians/about). The American Statistical Association has done an admirable job on its profiles of the life and works of the great statisticians. Note that adherence to the Fisher test is *not* employed by the association as a criterion of inclusion.

13. Mayo 1989, xii. Cf. Student 1938; Student 1927; and Jeffreys 1939, 8–9, 13–14.

14. Cf. Lehmann 1993, which comes to conclusions similar to Mayo's while telling a similarly erroneous history of modern significance testing.

15. Gigerenzer et al. 1989, 206.

16. See, for example, Arrow 1962; and Cyert and DeGroot 1987 and references there.

17. Feynman quoted in S. J. Press 2003, 230; Zellner 2005a, 7–8.

18. Feyerabend 1975, 20–1; italics in original.

Chapter 14

1. See E. S. Pearson 1990, 26–27.

2. K. P., ca. 1905, "Dice," 10–15, Box 67, Karl file, UCL; italics in original.

3. Fisher 1922, 1926b, 1956; cf. Zabell 1989.

4. E. Pearson 1966, 9.

5. E. Pearson 1990, 110.

6. Goodman 1999a, 995; see also 1992, 1993, and 1999b.

Chapter 15

1. Mosteller to Ziliak and McCloskey, May 21, 2005; Rothman to Ziliak, January 30, 2006.

2. http://www.stat.duke.edu/~berger.

3. Berger 2003, 4; Sterne and Davey Smith 2001.

4. Fidler et al. 2004a, 609; Thompson 1996, 1997.

5. Fidler et al. 2004b, 121.

6. *APA Manual* 2001, 22.

7. Rothman 1998, quoted in Fidler et al. 2004b, 121.

8. Rothman to Ziliak, e-mail communication, January 27, 2006.

9. Fidler et al. 2004b, 122.

CHAPTER 16

1. Klein, Elifson, and Sterk 2003, 167.
2. Rothman to Ziliak, personal interview, January 30, 2006.
3. Luby, Jones, and Horan 1993, 32.
4. Ibid., 38.
5. Paul Feyerabend was characteristically interesting on the point: a pluralistic conception of both science and society, he said, will consider evidence of all kinds, including the introspection of herbalists. See, for example, his "Notes on Relativism" (Feyerabend 1988).
6. Ziliak (with Hannon) 2006.

CHAPTER 17

1. Edgeworth 1885, 1896, 1907.

CHAPTER 18

1. Stigler 2005b, 63; 2000.
2. Scott 1953, 20; Stigler 1986, chaps. 1, 4.
3. Porter 2004, passim; E. Pearson 1936, 197–207; Pearson Papers, Karl file, passim, UCL.
4. In E. Pearson 1936, 194–99, 225.
5. In E. Pearson 1936, 211, 248.
6. Because of Galton's failing health Pearson had served since 1907 as director of the Galton Laboratory for National Eugenics, a duty he retained as the second Galton Professor of Eugenics from 1911 to 1933. Upon his retirement, Fisher would succeed him.
7. Quoted in Porter 2004, 7.
8. Box 1978, 49–52, 60–61. Incidentally, Leonard Darwin, son of Charles, was important to Gower Street in other ways, such as for publicizing eugenic ideas, if sometimes too loosely for Pearson's and Galton's taste, and for saving a youthful and unfocused R. A. Fisher from poverty and obscurity.
9. Folder 230, Pearson Papers, Karl file, UCL; E. Pearson 1938, 177–78, 217. See, for example, Pearson and Moul 1925, 1927; and Pearson and Lee 1903. Interestingly, some of the skulls examined by the Biometric Lab were a gift to Pearson from the great anthropologist Franz Boas, antiracist though Boas was (Porter 2004, 253). Also see Gould 1981; and Kevles 1985.
10. K. Pearson, ca. 1905, "On the Fundamental Problem of Statistics," 2, Pearson Papers, Box 67.
11. Galton, 1899, "The Risk of Misclassification When the Objects Classified Vary Continuously and the Examiners Are Fallible," Galton Papers, Box 156, 6.

CHAPTER 19

1. K. Pearson, ca. 1908, "Variates," 2, Box 67, Karl file, Pearson Papers, UCL.
2. E. Pearson 1990, 14; see also E. Pearson 1936, 1938.

CHAPTER 20

1. In E. S. Pearson 1938, 182–87; *K. P. Lectures Vol. I,* 1906, Gosset file, Pearson Papers, UCL.

2. Student 1908a; E. Pearson 1990, 47–48.

3. Fisher 1915a; Student 1925; E. Pearson 1968, 446–48; E. Pearson 1990, 49.

4. Egon planned to write a book on Gosset's statistical ideas and scientific contributions. He also intended the book to convey some of Gosset's personal details and family history. The book was incomplete at the time of Egon's death in 1980. A short biography of Gosset (Pearson 1990) was assembled in 1990 by R. L. Plackett who, with the assistance of G. A. Barnard, edited chiefly out of Egon's unpublished notes, correspondence, and published articles.

5. Cf. Gosset to Fisher, 1918, Letters 1–2, in Gosset 1962.

6. Folders 282–83, Pearson Papers, Gosset file, UCL; Student 1908a, 13–20.

7. McMullen 1939, 208, 209–10.

8. Fisher 1939; McMullen 1939, 205–10; "H. H."[Harold Hotelling] 1938b, 249; E. Pearson 1939, 211; Boland 1984; "E. M. E.," in "J. W." [John Wishart] 1938, 251; Reid, 150; B. Gosset, in E. Pearson 1990, 21.

9. Gosset, in E. Pearson 1968, 446.

10. Gosset 1962, Letter 2.

11. Ibid., Letter 2, 6; italics supplied.

12. Gosset, Folder 283, Pearson Papers, UCL.

13. Letter, W. S. Gosset to K. Pearson, March 29, 1932, Pearson Papers, Gosset file, UCL; italics supplied.

14. Italics supplied.

15. For example, see Letters 69, 76, 159, in Gosset 1962; B. Gosset, in E. Pearson 1990, 21; and Ziliak 2008b.

16. Gosset, 1918, Letter to K. P., in E. Pearson 1990, 39.

17. Lindley 1991.

18. Folder 282, "Student's Haemocytometer Paper on Yeast-Cell Counts," 1905–1907, UCL; Student 1907. Dennison and MacDonagh (1998) find evidence that the pseudonyms "Pupil" and "Student" were proposed by Gosset's managing director, C. D. La Touche, when the question of publishing Gosset's yeast cell paper came up (90, n. 9; cf. "Research Papers," October 29, 1906, Guinness Archives, GDB/C004.09/0004.14). Our point is that Gosset began the "yeast cell" paper in 1905, recording his data and writing down his ideas for it in *The Student's Science Notebook,* of which Dennison and MacDonagh were unaware.

CHAPTER 21

1. Savage 1971a; Kruskal 1980.

2. Fisher, in Gosset 1936, 122–23; Box 1978, chaps. 1–2.

3. Letters 20, 23, in Gosset 1962.

4. Letters 29, 30, in Gosset 1962; Gosset 1923, in E. S. Pearson 1990 [posthumous], 58.

5. If Gosset ever lost his patience with Fisher with regard to the economic design of experiments, it was in Student's crushingly persuasive "Comparison between Balanced and Random Arrangements of Field Plots" (Student 1938).

6. Letter 29, July 25, 1923, in Gosset 1962.

7. Pearson 1990, 59.

8. Letters 101–8, in Gosset 1962.

9. Gosset himself had been writing a textbook on experimental design and evaluation since 1910. Brewery work and other duties forever got in the way. The textbook long in process was accidentally destroyed in 1934 when he and his family moved from Dublin to London (E. Pearson 1990, 19).

10. Fisher 1935, 34–44, 191–93.

11. Dennison and MacDonagh 1998, 89; J. Fisher Box 1987, 46; Lynch and Vaizey 1960; Matthias 1959, 166–67.

12. Beaven 1947, ix, 57, 274–93.

13. Fisher 1955, 69, 70, 74; Fisher 1945, in Bennett, ed., 1990, 2, 9, 23, and throughout; 1956, 98–104.

14. Neyman to Fisher, June 13, 1933, and Fisher to Neyman, June 17, 1933, both in Bennett 1990, 192.

15. Good matches for Box's 1978 biography of her father are Kruskal's "Significance of Fisher" (1980) and Savage's "On Re-reading R. A. Fisher" (1971a).

16. Box 1997, 105.

17. See Gosset 1936, but especially Student 1938. See also Neyman and Pearson 1938; and Jeffreys 1939b. Recall from chapter 3 the dozen or so standard defenses of status quo significance testing. The same defenses, plus a few others, are typically heard in reference to Fisher's "randomized" designs of experiments. But Student (1938) shows Fisher's "randomized" designs to be both less *precise* and less *economically relevant* than are Gosset's "balanced" designs.

18. Neyman 1961, 148–49, quoted in Savage 1971a, 446.

19. Gutteling et al. 2007, 31–32.

20. Letter, W. S. Gosset to K. Pearson, July 14, 1931, Folder 525, UCL. Gosset closed the letter thus: "Would you care to have a look at this [note] when written, with a view to publication? I think that criticism of method is more appropriate for publication in Biometrika than anywhere else." Karl Pearson, too, could be an imperious man, but he used his journal of eugenics and biometry to encourage debates and advances in statistical methods. Though disagreeing vehemently with their methods, K. P. published Gosset's and Egon's and Neyman's career-making papers.

21. Fisher worried in *Statistical Methods and Scientific Inference* (1956, 99) about the relationship between randomness and tests of significance. They had "no useful part to play," he wrote (99). In *Statistical Methods for Research Workers* (1925), as in practice, he acted on this faith. The journals of, for example, economics, forensics, education, and management science are filled with tests on nonrandom samples of convenience.

Chapter 22

1. Pearson 1990, 49–53; Letter, Fisher to Dufrenoy, May 1936, in Bennett 1990, 308.

2. Letter, Gosset to Fisher, cited in Box 1978, 117.

3. Letter 11, in Gosset 1962; cf. E. Pearson 1990, 49.

4. Gosset to Fisher, Letter 26, in Gosset 1962.

5. Letter 47, in Gosset 1962.

6. E. Pearson 1990, 51.

7. Letter 48, in Gosset 1962.

8. Letter 66, in Gosset 1962..

9. For example, one was from J. Dunlap, a Stanford psychometrician, in 1927: Letter 86, in Gosset 1962.

10. Letter, Fisher to Deming, September 25, 1934, in Bennett 1990, 80–82.

11. Letter, Fisher to S. S. Wilks, February 6, 1934, in Bennett 1990, 302–4.

12. An exception might be Hotelling's "On the Generalization of Student's Ratio" (1931). But in this, too, Hotelling's paper on T^2, the spirit is thoroughly Fisherian.

13. In an obscure letter of May 12, 1936, to H. D. Dufrenoy, a French scientist, Fisher went some way to correct a confused Dufrenoy, who believed erroneously with the rest of the world that the test and table were Fisher creations (Fisher to H. D. Dufrenoy, May 12, 1936, in Bennett 1990, 307–8). One unpublished letter to a French pathologist cannot make up for the world of confusion Fisher created. And even then Fisher did not go far enough: for example, he did not tell Dufrenoy that Fisher himself had erased the economic and Bayesian elements of Gosset's approach nor did he anywhere indicate that Gosset was a lifelong opponent of Fisher's own 5 percent rule. The question is why Fisher "came out" at all to Mr. Dufrenoy.

14. Edgeworth 1896, 358.

15. Gigerenzer 2004, 587, for instance.

16. Fisher 1956, 100–104; 1955, passim.

17. Letter, R. A. Fisher to A. T. James, March 13, 1962, in Bennett 1990, 148; italics supplied.

18. Darnell, 29–35. See also Mirowski and Hands 1998; and Hands and Mirowski 1998. Confusion in the 1920s about statistical significance led to confusion in the 1930s about demand functions.

19. Darnell, 7, n. 26.

20. "The only probability notion I can make sense of," Friedman said in an interview in 1996 at the annual meetings of the American Economic Association in San Francisco, "is personal probability in the spirit of Savage and others. Keynes's degree of belief is in the same family," Friedman continued. "In fact I believe," he said, "that Keynes's contribution in his *Probability* book has been underrated and overlooked" (interview in Snowdon and Vane 1999). We would add that the probabilistic philosophies of Gosset, Frank Ramsey, and Bruno de Finetti are also in the same family and have been overlooked. Savage (1954) acknowledges the similarity of his own approach to that of Ramsey (279) but appears here and in all his writings to be innocent of Gosset's personalism, developed fifty years before. Contrary to Friedman's comparison, Savage, it would seem, had rejected any exact comparison of his own approach with that of Keynes's *Treatise* (39, 56, 277).

21. A. Fisher 1991, 310.

22. Letter, R. Fisher to Arne Fisher, in Bennett 1990, 313.

23. See, for example, Neyman 1960, 8–9.
24. R. Fisher 1951, 145.

CHAPTER 23

1. "Signifying Nothing?" *Economist*, January 31, 2004; "Taking on 'Rational Man,'" *Chronicle of Higher Education*, January 24, 2003.

2. See, for example, Yates and Mather 1963, 93; Dawkins 1986, 163–64; and Dawkins 1982, 137, 238–39. Ernst Mayr (1982, 553–55), the great historian of biology, doubted that Fisher was the first scientist to successfully meld Mendelian inheritance with biometrics and evolution by natural selection. "We still lack a good history of this period," he said (553). Letters from Gosset to E. S. Beaven, 1908 to 1914, suggest that Gosset may have gotten *there* first, too (Ziliak 2008a; Beaven Collection, Museum of English Rural Life).

3. By contrast, most historians of science no longer believe in a tidy and Whiggish progress of "exactness" over time. But in 1959 even some leading historians and scientists did. A conference on "The History of Quantification in the Sciences," which included Bernard Cohen, Thomas Kuhn, Bernard Barber, Samuel Wilks, Edwin Boring, Simon Kuznets, Harry Woolf, and Joseph Spengler—and whose papers were published in a special issue of *Isis* (52[2] [June 1961])—used the words *exact* and *precise* as if the status and definition of those words required no critical attention. Precise, they seemed to say, just *was* better. Perhaps Friedman and Kruskal and company were not in Chicago or able to attend the conference.

CHAPTER 24

1. Kuhn 1962, 46, 174–91.

2. Gigerenzer et al. 1989, 101–2 (Gigerenzer 2004, by contrast, gets straight to the point); MacKenzie 1981, 111–19.

3. Efron's favorable assessment of Fisher has been favorably cited in turn by others and notably by medical researchers in Australia. "[S]ignificance tests are 'alive and well,'" the authors conclude their article on the history of p-values (Moran and Soloman 2004, 134). One of the authors is stationed in Adelaide—Fisher's last stand and resting place.

Works Cited

Archival Sources

Beaven Collection, Museum of English Rural Life, University of Reading.

Galton Papers, University College London (UCL), Special Collections.

Gosset Collection, Guinness Archives, Guinness Storehouse, Dublin (Diageo).

Letters of William Sealy Gosset to R. A. Fisher, Vols. 1–5. 1962. Eckhart Library, University of Chicago. In 1957 Fisher gave his collection of letters from Gosset to a mutual friend and brewer, Launce McMullen, who, in 1962, the year of Fisher's death, printed it for private circulation.

Pearson Papers (Karl Pearson, Egon Pearson, Gosset, Fisher, Neyman, and Shaw files), UCL Special Collections.

Published Material

Adkins, Dorothy C. 1968. L. L. Thurstone. In *International Encyclopedia of the Social Sciences*, ed. D. Sills, 22–25.

Agee, James, and Walker Evans. 1941. *Let Us Now Praise Famous Men.* New York: Mariner.

Altman, Douglas G. 1991. Statistics in Medical Journals: Developments in the 1980s. *Statistics in Medicine* 10:1897–1913.

Altman, Douglas G. 2000. Confidence Intervals in Practice. In *Statistics with Confidence*, ed. D. G. Altman, D. Machin, T. N. Bryant, and M. J. Gardner, 6–14. London: BMJ Books.

Altman, Morris. 2004. Statistical Significance, Path Dependency, and the Culture of Journal Publication. *Journal of Socio-Economics* 33(5): 651–63.

Altman, Morris, ed. 2006. *Handbook of Contemporary Behavioral Economics.* Armonk, NY: M. E. Sharpe.

Amemiya, Takeshi. 1985. *Advanced Econometrics.* Cambridge: Harvard University Press.

American Economic Review. 1980–99. The 369 full-length articles published between January 1980 and December 1999 that use tests of statistical significance, May supplement excluded.

American Psychological Association. 1952–2001 [revisions]. Editions of the *Publication Manual of the American Psychological Association* published between 1952 and 2001. Washington, DC: American Psychological Association.

Anderson, D. R., K. P. Burnham, and W. Thompson. 2000. Null Hypothesis Testing: Problems, Prevalence, and an Alternative. *Journal of Wildlife Management* 64:912–23.

Angrist, Joseph. 1995. The Economic Returns to Schooling in the West Bank and Gaza Strip. *American Economic Review* 85(5): 1065–86.

Arrow, Kenneth J. 1959 [1960]. Decision Theory and the Choice of a Level of Significance for the *t*-Test. In Olkin et al., *Contributions,* 70–78.

Arrow, Kenneth J. 1962. The Economic Implications of Learning by Doing. *Review of Economic Studies* 29(3): 155–73.

Arrow, Kenneth J., David Blackwell, and M. A. Girshick. 1949. Bayes and Minimax Solutions of Sequential Decision Problems. *Econometrica* 17:213–44.

Atkins, L., and D. Jarrett. 1979. The Significance of "Significance Tests." In *Demystifying Social Statistics,* ed. J. Irvine, I. Miles, and J. Evans. London: Pluto.

Atkinson, A. B. 1999. *The Economic Consequences of Rolling Back the Welfare State.* Cambridge: MIT Press.

Atlas, James. 2002. *Bellow.* New York: Random House.

Baird, Davis. 1988. Significance Tests, History, and Logic. In *Encyclopedia of Statistical Sciences,* ed. S. Kotz and N. L. Johnson, 8:466–71. New York: Wiley.

Bakan, David. 1966. The Test of Significance in Psychological Research. *Psychological Bulletin* 66(6): 423–37.

Banks, David. 1996. A Conversation with I. J. Good. *Statistical Science* 11(1): 1–19.

Batterham, A. M., and W. G. Hopkins. 2005. A Decision Tree for Controlled Trials. *SportScience* 9 (December): 33–39.

Batterham, Alan M., and William G. Hopkins. 2005. Making Meaningful Inferences about Magnitudes. *International Journal of Sports Physiology and Performance* 1:50–57.

Beaven, E. S. 1947. *Barley: Fifty Years of Observation and Experiment.* Ed. N. Stallwood. London: Duckworth.

Becker, Gary S., Michael Grossman, and Kevin M. Murphy. 1994. An Empirical Analysis of Cigarette Addiction. *American Economic Review* 84(3): 396–418.

Beeton, Mary, and Karl Pearson. 1901. On the Inheritance of the Duration of Life and on the Intensity of Natural Selection in Man. *Biometrika* 1(1): 50–89.

Bellow, Saul. 1944 [1977]. *The Dangling Man.* New York: Penguin.

Bennett, J. H., ed. 1990. *Statistical Inference and Analysis: Selected Correspondence of R. A. Fisher.* Oxford: Clarendon.

Berenson, Alex. 2005. Newly Disclosed e-mails add Vioxx Wrinkle. *Chicago Tribune,* April 24, A14.

Berg, Nathan. 2004. No-Decision Classification: An Alternative to Testing for Statistical Significance. *Journal of Socio-Economics* 33(5): 631–50.

Berger, James O. 2003. Could Fisher, Jeffreys, and Neyman Have Agreed on Testing? *Statistical Science* 18(1): 1–32.

Berger, James O., and Robert L. Wolpert. 1984 [1988]. *The Likelihood Principle,* 2nd ed. Hayward, CA: Institute of Mathematical Statistics.

Bernando, Jose M. 2006. The Valencia Story: Some Details on the Origin and Development of the Valencia International Meetings on Bayesian Statistics. Universitat de Valencia, Spain.

Bernanke, Ben S., and Alan S. Blinder. 1992. The Federal Funds Rate and the Channels of Monetary Transmission. *American Economic Review* 82(4): 901–21.

Bernheim, B. Douglas, and Adam Wantz. 1995. A Tax-Based Test of the Dividend Signaling Hypothesis. *American Economic Review* 85(3): 532–51.

Billig, Michael. 1999. *Freudian Repression: Conversation Creating the Unconscious*. Cambridge: Cambridge University Press.

Biometrika. 1902.1(1): 164–76.

Biometrika, 1901–34. Every article published between 1901 and 1934 that uses a test of statistical significance.

Black, Max. 1970. *Margins of Precision: Essays in Logic and Language*. Ithaca: Cornell University Press.

Black, Richard. 2005. Japanese Whaling "Science" Rapped. *BBC News World Edition*, June 22, 2005. http://news.bbc.co.uk/2/hi/science/nature/

Blackwell, David, and M. A. Girshick. 1954. *The Theory of Games and Statistical Decisions*. New York: Wiley.

Bloomquist, Glenn C., Mark C. Berger, and John P. Hoehn. 1988. New Estimates of Quality of Life in Urban Areas. *American Economic Review* 78(1): 89–107.

Boland, Philip J. 1984. A Biographical Glimpse of William Sealy Gosset. *American Statistician* 38(3): 179–83.

Boring, Edwin G. 1919. Mathematical versus Scientific Significance. *Psychological Bulletin* 16(10): 335–38.

Boring, Edwin G. 1929. *A History of Experimental Psychology*. London: Century.

Bossier, M., J. B. Knight, and R. H. Sabot. 1985. Earnings, Schooling, Ability, and Cognitive Skills. *American Economic Review* 75(3): 1016–30.

Bowen, M. E., and J. A. Mazzco, eds. 1979. *Writing about Science*. Oxford: Oxford Univesity Press.

Box, Joan Fisher. 1978. *R. A. Fisher: The Life of a Scientist*. New York: Wiley

Box, Joan Fisher. 1987. Guinness, Gosset, Fisher, and Small Samples. *Statistical Science* 2 (February): 45–52.

Box, Joan Fisher. 1997. Fisher, Ronald A. In Johnson and Kotz, *Leading Personalities in Statistical Sciences*, 99–108.

Bradburn, Norman. 2005. "William Kruskal." Memorial Service for William Kruskal, May 21, 2005, University of Chicago.

Burke, Kenneth. 1935. *Permanence and Change*. New York: New Republic.

Burke, Kenneth. 1946. Letter to David Cox, August 14, 1946, http://www.lib.siu.edu/spcol/inventory/part1.htm.

Burnand, B., W. N. Kernan, and A. R. Feinstein. 1990. Indexes and Boundaries for "Quantitative Significance" in Statistical Decisions. *Journal of Clinical Epidemiology* 43:1273–84.

Cain, Glen G., and Harold W. Watts. 1970. Problems in Making Policy Inferences from the Coleman Report. *American Sociological Review* 35(2): 228–42.

Cairncross, Alec. 1992. From Theory to Policy-Making: Economics as a Profession. *Banco Nazionale del Lavoro Quarterly Review* 180 (March): 3–20.

Card, David, and Alan B. Krueger. 1994a. Minimum Wages and Employment: A Case Study of the Fast-Food Industry in New Jersey and Pennsylvania. *American Economic Review* 84(2): 772–93.

Card, David, and Alan B. Krueger. 1994b. *Myth and Measurement: The New Economics of the Minimum Wage.* Princeton: Princeton University Press.

Carver, Ronald P. 1978. The Case against Statistical Significance Testing. *Harvard Educational Review* 48(3): 378–98.

Carver, Ronald P. 1993. The Case against Statistical Significance Testing Revisited. *Journal of Experimental Education* 61:287–92.

Chernoff, Herman. 1968 [1978]. Decision Theory. In Kruskal and Tanur, *International Encyclopedia of Statistics,* 131–35.

Cicero, Marcus Tullius. 45 BC [1938]. *De Dininatione,* In *De Senectute, de Amicitia, de Divinatione,* ed. and trans. W. A. Falconer. Cambridge: Harvard University Press.

Clark, Kim B. 1984. Unionization and Firm Performance. *American Economic Review.* 74(5): 893–19.

Cohen, I. B. 2005 [posthumous]. *The Triumph of Numbers.* New York: Norton.

Cohen, Jacob. 1962. The Statistical Power of Abnormal-Social Psychological Research: A Review. *Journal of Abnormal and Social Psychology* 65(3): 145–53.

Cohen, Jacob. 1969. *Statistical Power Analysis for the Behavioral Sciences.* Hillsdale: Elbaum Associates.

Cohen, Jacob. 1994. The Earth Is Round ($p < 0.05$). *American Psychologist* 49:997–1003.

Cohn, Jay. 1981. Multicenter Double-Blind Efficacy and Safety Study Comparing Alprazolam, Diazepam, and Placebo in Clinically Anxious Patients. *Journal of Clinical Psychiatry* 9:347–51.

Colander, David. 2000. New Millennium Economics: How Did It Get This Way and What Way Is It? *Journal of Economic Perspectives* 14(1): 121–32.

Cole, Russell, and Graham McBride. 2004. Assessing Impacts of Dredge Spoil Disposal Using Equivalence Tests: Implications of a Precautionary (Proof of Safety) Approach. *Marine Ecology Progress Series* 279:63–72.

Csada, R. D., P. C. James, and R. H. M. Espie. 1996. The "File Drawer Problem" of Non-significant Results: Does It Apply to Biological Research? *Oikos* 76:591–93.

Cumming, G., and S. Finch. 2005. Inference by Eye: Confidence Intervals and How to Read Pictures of Data. *American Psychologist.* 60: 170–80.

Cutler, S. J., and S. W. Greenhouse, J. Cornfield, and M. A. Schneiderman. 1966. The Role of Hypothesis Testing in Clinical Trials. *Journal of Chronic Diseases* 19: 857–92.

Cyert, Richard M., and Morris H. DeGroot. 1987. *Bayesian Analysis and Uncertainty in Economic Theory.* Totowa, NJ: Rowman and Littlefield.

Darby, Michael R. 1984. The U.S. Productivity Slowdown. *American Economic Review* 74(3): 301–22.

Darnell, A. C. 1997. Imprecise Tests and Imprecise Hypotheses. *Scottish Journal of Political Economy* 44(3): 247–68.

David, Florence N., ed. 1966. *Research Papers in Statistics: Festschrift for J. Neyman.* London: Wiley.

David, Herbert A. 1981. In Memoriam: Egon S. Pearson, 1895–1980. *American Statistician* 35(2): 94–95.

Dawkins, Richard. 1982 [1999]. *The Extended Phenotype.* Oxford: Oxford University Press.

Dawkins, Richard. 1986 [1996]. *The Blind Watchmaker.* New York: Norton.

De Finetti, Bruno. 1971 [1976]. Comments on Savage's "On Re-reading R. A. Fisher." *Annals of Statistics* 4(3): 486–87.

DeGroot, Morris H. 1975 [1989]. *Probability and Statistics.* Reading, MA: Addison-Wesley.

Deming, W. Edwards. 1938 [1943]. *Statistical Adjustment of Data.* New York: Dover.

Deming, W. Edwards. 1950. *Some Theory of Sampling.* New York: Wiley.

Deming, W. Edwards. 1961. *Sample Design in Business Research.* New York: Wiley.

Deming, W. Edwards. 1975. On Probability as a Basis for Action. *American Statistician* 29(4): 146–52.

Deming, W. Edwards. 1982. *Out of the Crisis.* Cambridge: Center for Advanced Engineering Study, Massachusetts Institute of Technology.

Dennison, S. R., and Oliver MacDonagh. 1998. *Guinness, 1886–1939.* Cork: Cork University Press.

Dewey, John. 1929. *The Quest for Certainty.* New York: Minton, Balch.

Dillard, Annie. 1974. *Pilgrim at Tinker Creek.* New York: Harper and Row.

Dillard, Annie. 1988. *Living by Fiction.* New York: Harper and Row.

Douglass, T. J., and L. D. Fredendall. 2004. Evaluating the Deming Management Model. *Decision Sciences* 35(3): 325–32.

Edgeworth, Francis Y. 1881. *Mathematical Psychics.* London: C. K. Paul.

Edgeworth, Francis Y. 1885. Methods of Statistics. In *Jubilee Volume of the Statistical Society,* 181–217. London: Royal Statistical Society of Britain.

Edgeworth, Francis Y. 1896. Statistics of Unprogressive Communities. *Journal of the Royal Statistical Society* 5(2): 358–86.

Edgeworth, Francis Y. 1907. Statistical Observations on Wasps and Bees. *Biometrika* 5(4): 365–86.

Edwards, A. W. F. [no date] R. A. Fisher on Karl Pearson. *Notes Rec. R. Soc. Lond.* 48(1): 97–106.

Edwards, W., H. Lindman, and L. J. Savage. 1963. Bayesian Statistical Inference for Psychological Research. *Psychological Review* 70(3): 193–242.

Efron, Bradley. 1996 [1998]. R. A. Fisher in the 21st Century. *Statistical Science* 13(2): 95–122.

Eisenberger, Mario A., et al. 1998. Bilateral Orchiectomy with or without Flutamide for Metastatic Prostate Cancer. *New England Journal of Medicine* 339:1036–42.

Elashoff, Janet D., and Robert M. Elashoff. 1968 [1978]. Effects of Errors in Statistical Assumptions. In Kruskal and Tanur, *International Encyclopedia of Statistics,* 1:229–50.

Elliott, Graham, and Clive W. J. Granger. 2004. Evaluating Significance: Comments on "Size Matters." *Journal of Socio-Economics* 33(5): 547–50.

Emerson, J. D., and G. A. Colditz. 1992. Use of Statistical Analysis. In *New England Journal of Medicine.*

Epidemiology. 1990–2001. Every "original article" published between January 1990 and December 2001 that uses quantitative methods. In *Medical Uses of Statistics,* ed. J. C. Bailar and F. Mosteller, 45–57. Boston: NEJM Books.

Felson, D. T., L. A. Cupples, and R. F. Meenan. 1984. Misuse of Statistical

Methods in *Arthritis and Rheumatism,* 1982 versus 1967–68. *Arthritis and Rheumatism* 27:1018–22.

Feyerabend, Paul K. 1975 [1993]. *Against Method.* London: Verso.

Feyerabend, Paul K. 1987. *Farewell to Reason.* New York: Verso.

Feynman, Richard P. 1996. *Feynman Lectures on Computation.* Ed. A. J. G. Hey and R. W. Allen. Reading, MA: Perseus Books.

Feynman, Richard P., Robert B. Leighton, and Matthew Sands. 1965. *The Feynman Lectures on Physics.* 3 vols. Reading, MA: Addison-Wesley.

Fidler, Fiona. 2002. The Fifth Edition of the *APA Publication Manual:* Why Its Statistics Recommendations Are So Controversial. *Educational and Psychological Measurement* 62:749–70.

Fidler, Fiona, Geoff Cumming, Mark Burgman, and Neil Thomason. 2004a. Statistical Reform in Medicine, Psychology, and Ecology. *Journal of Socio-Economics* 33(5): 615–30.

Fidler, Fiona, N. Thomason, G. Cumming, S. Finch, and J. Leeman. 2004b. Editors Can Lead Researchers to Confidence Intervals, but They Can't Make Them Think: Statistical Reform Lessons from Medicine. *Psychological Science* 15:119–26.

Fienberg, Stephen E., and Arnold Zellner, eds. 1975. *Studies in Bayesian Econometrics in Honor of Leonard J. Savage.* Amsterdam: North-Holland.

Fisher, Arne. 1931. A. Fisher to R. A. Fisher. In Bennett, *Statistical Inference and Analysis,* 310–12.

Fisher, R. A. 1914. Some Hopes of a Eugenicist. *Eugenics Review* 5:309–15.

Fisher, R. A. 1915a. Frequency Distribution of the Values of the Correlation Coefficients in Samples from an Indefinitely Large Population. *Biometrika* 10: 507–21.

Fisher, R. A. 1915b. Racial Repair. *Eugenics Review* 7:204–7.

Fisher, R. A. 1921. On the Probable Error of a Correlation Coefficient Deduced from a Small Sample. *Metron* 1:3–32.

Fisher, R. A. 1922. On the Mathematical Foundations of Theoretical Statistics. *Philos. Trans. Roy. Soc. London,* ser. A, 222:309–68.

Fisher, R. A. 1923. Statistical Tests of Agreement between Observation and Hypothesis. *Econometrica* 3:139–47.

Fisher, R. A. 1925a [1941]. *Statistical Methods for Research Workers.* New York: G. E. Stechart. The eighth of thirteen editions published in at least seven languages. Originally published in Edinburgh by Oliver and Boyd.

Fisher, R. A. 1925b. Applications of "Student's" Distribution. *Metron* 5(3): 90–104.

Fisher, R. A. 1925c. Expansion of "Student's" Integral in Powers of n^{-1}. *Metron* 5(3): 110–20.

Fisher, R. A. 1926a. Arrangement of Field Experiments. *Journal of Ministry of Agriculture* 23:503–13.

Fisher, R. A. 1926b. Bayes' Theorem. *Eugenics Review* 18:32–33.

Fisher, R. A. 1931. Letter to Arne Fisher. In Bennett, *Statistical Inference and Analysis,* 313.

Fisher, R. A. 1932. Family Allowances in the Contemporary Economic Situation. *Eugenics Review* 24:87–95.

Fisher, R. A. 1935. *The Design of Experiments.* Edinburgh: Oliver and Boyd. The first of eight editions in at least four languages.

Fisher, R. A. 1936. The Significance of Regression Coefficients. *Colorado College Publ. Gen. Ser.* 208:63–67. Abstract, Cowles Commission Research Conference on Economics and Statistics.

Fisher, R. A. 1939. Student. *Annals of Eugenics.* 9:1–9.

Fisher, R. A. 1950. *Contributions to Mathematical Statistics.* Ed. Walter A. Shewhart. New York: Wiley.

Fisher, R. A. 1951. Letter to W. E. Hick. In Bennett, *Statistical Inference and Analysis,* 145.

Fisher, R. A. 1955. Statistical Methods and Scientific Induction. *Journal of the Royal Statistical Society.* Ser. B (Methodological) 17(1): 69–78.

Fisher, R. A. 1956 [1959]. *Statistical Methods and Scientific Inference.* 2nd ed. New York: Hafner.

Fisher, R. A., and Frank Yates. 1938 [1963]. *Statistical Tables for Biological, Agricultural, and Medical Research.* 6th ed. Edinburgh: Oliver and Boyd.

Fleiss, J. L. 1986. Significance Tests Do Have a Role in Epidemiological Research: Reaction to A. A. Walker. *American Journal of Public Health* 76:559–60.

Florens, Jean-Pierre, and Michel Mouchart. 1993. Bayesian Testing and Testing Bayesians. In Maddala, Rao, and Vinod, *Handbook of Statistics,* 11:303–91.

Fraley, R. Chris. 2003. The Statistical Significance Testing Controversy: A Critical Analysis. Course syllabus for PSCH 548, Graduate Seminar in Methods and Measurement, Department of Psychology, University of Illinois, Chicago. http://www.uic.edu/classes/psych/psych548/fraley/.

Fraser, D. A. S. 1958. *Statistics: An Introduction.* New York: Wiley.

Freedman, David, Robert Pisani, Roger Purvis. 1978. *Statistics.* New York: Norton.

Freiman, Jennie A., T. Chalmers, H. Smith, and R. R. Kuebler. 1978. The Importance of Beta, the Type II Error, and Sample Design in the Design and Interpretation of the Randomized Control Trial: Survey of 71 "Negative" Trials. *New England Journal of Medicine* 299:690–94.

Frenkel, Jacob. 1978. Purchasing Power Parity: Doctrinal Perspectives and Evidence from the 1920s. *Journal of International Economics* 8 (May): 169–91.

Friedman, Milton. 1991. A Cautionary Tale about Multiple Regression (the Appendix to "Alternative Approaches to Analyzing Economic Data," by Milton Friedman and Anna J. Schwartz). *American Economic Review* 81(1): 48–49. Reprinted from M. Friedman, 77–92. In W. Breit and R. W. Spencer, eds., *Lives of the Laureates.* Cambridge: MIT Press, 1990.

Friedman, Milton. 1996. Milton Friedman, In *Conversations with Leading Economists,* ed. B. Snowdon and H. Vane, 124–44. Elgar, 1999. Interview.

Frisch, Ragnar. 1934. *Statistical Confluence Analysis by Means of Complete Regression Systems.* Oslo: University Institute of Economics.

Gabor, Andrea. 1992. *Deming: The Story of How W. Edwards Deming Brought the Revolution to America.* New York: Penguin.

Gabor, Andrea. 2006. Deirdre McCloskey's Market Path to Virtue. *Strategy + Business* 43 (summer).

Galton, Francis. 1872. "Statistical Inquiries into the Efficacy of Prayer." *Fortnightly Review* 12:125–35.

Gelman, Andrew, and Hal Stern. 2006. The Difference between "Significant" and "Not Significant" Is Not Itself "Statistically Significant." *American Statistician* 60(4): 328–31.

Giere, Ronald N. 1979. Foundations of Probability and Statistical Inference. In *Current Research in Philosophy of Science*, ed. Peter D. Asquith and Henry E. Kyburg Jr., 503–33. East Lansing, MI: Philosophy of Science Association.

Giere, Ronald N., and Richard S. Westfall, eds. 1973. *Foundations of Scientific Method: The Nineteenth Century*. Bloomington: Indiana University Press.

Gigerenzer, Gerd. 2004. Mindless Statistics. *Journal of Socio-Economics* 33(5): 587–606.

Gigerenzer, Gerd, Zeno Swijtink, Theodore Porter, Lorraine Daston, John Beatty, and Lorenz Kruger. 1989. *The Empire of Chance*. Cambridge: Cambridge University Press.

Girshick, M. A., and Leonard J. Savage. 1951. Bayes and Minimax Estimates for Quadratic Loss Functions. In *Proceedings of the Second Berkeley Symposium on Mathematical Statistics and Probability*, ed. J. Neyman, 53–74. Berkeley: University of California Press.

Goldberger, Arthur S. 1991. *A Course in Econometrics*. Cambridge: Harvard University Press.

Good, I. J. 1981. Some Logic and History of Hypothesis Testing. In *Philosophy in Economics*, ed. J. C. Pitt. Dordrecht: Reidel.

Good, I. J. 1992. The Bayes/Non-Bayes Compromise: A Brief Review. *Journal of the American Statistical Association* 87:597–606.

Goodman, S. 1992. A Comment on Replication, *p*-Values, and Evidence. *Statistics in Medicine* 11:875–79.

Goodman, S. 1993. *P*-Values, Hypothesis Tests, and Likelihood: Implications for Epidemiology of a Neglected Historical Debate. *American Journal of Epidemiology* 137:485–96.

Goodman, S. 1999a. Toward Evidence-Based Medical Statistics, part 1: The *p*-Value Fallacy. *Annals of Internal Medicine* 130:995–1004.

Goodman, S. 1999b. Toward Evidence-Based Medical Statistics, part 2: The Bayes Factor. *Annals of Internal Medicine* 130:995–1004.

Gosset, William S. 1936. Co-operation in Large-Scale Experiments. *Supplement to the Journal of the Royal Statistical Society* 3(2): 115–36. For additional works by Gosset, see the entries listed under Student, his pseudonym.

Gottfredson, Linda S. 1996. What Do We Know about Intelligence? *American Scholar* 65 (winter): 15–30.

Gould, Stephen Jay. 1981. *The Mismeasure of Man*. New York: Norton.

Granger, Clive W. J. 1994. A Review of Some Recent Textbooks of Econometrics. *Journal of Economic Literature* 32 (March): 115–22.

Graves, Robert, and Allen Hodge. 1943 [1979]. *The Reader over Your Shoulder*. New York: Viking.

Greene, Clinton A. 2003. Towards Economic Measures of Cointegration and Non-cointegration. Unpublished paper, Department of Economics, University of Missouri, St. Louis.

Greenspan, Alan. 2004. Risk and Uncertainty in Monetary Policy. *American Economic Review*, May supplement, 33–40.

Griliches, Zvi. 1986. Productivity, R & D, and Basic Research at the Firm Level in the 1970s. *American Economic Review* 76(1): 330–74.

Griliches, Zvi, and Michael D. Intriligator, eds. 1983–86. *Handbook of Econometrics*. 3 vols. Amsterdam: North-Holland.

Guilford, J. P. 1942. *Fundamental Statistics in Psychology and Education*. New York: McGraw-Hill.

Gutteling, E. W., J. A. G. Riksen, J. Bakker, and J. E. Kammenga. 2007. Mapping Phenotypic Plasticity and Genotype-Environment Interactions Affecting Life-History Traits in *Caenorhabditis Elegans*. *Heredity* 98:28–37.

Haavelmo, Trygve. 1944. The Probability Approach in Econometrics. *Econometrica* 12: iii–iv, 1–115.

Haavelmo, Trygve. 1958. The Role of the Econometrician in the Advancement of Economic Theory. *Econometrica* 26 (July): 351–57.

Hacking, Ian. 1975. *The Emergence of Probability*. Cambridge: Cambridge University Press.

Hacking, Ian. 1990 [1995]. *The Taming of Chance*. Cambridge: Cambridge University Press.

Hamermesh, Daniel S. 1989. Labor Demand and the Structure of Adjustment Costs. *American Economic Review* 79(4): 674–89.

Hamermesh, Daniel S. 1999. The Art of Labormetrics. Cambridge, MA: National Bureau of Economic Research.

Hands, D. Wade, and Philip Mirowski. 1998. Harold Hotelling and the Neoclassical Dream. In *Economics and Methodology: Crossing Boundaries*, ed. R. Backhouse, D. Hausman, U. Mäki, and A. Salanti, 322–97. New York: Macmillan and St. Martin's.

Harlow, Lisa Lavoie, Stanley A. Mulaik, and James H. Steiger, eds. 1997. *What If There Were No Significance Tests?* Mahwah, NJ: Lawrence Erlbaum Associates.

Hayek, Friedrich A. 1952. *The Counter-revolution of Science*. Glencoe, IL: Free Press.

Hayes, J. P., and R. J. Steidl. 1997. Statistical Power and Analysis and Amphibian Population Trends. *Conservation Biology* 11:273–75.

Heckman, James J. 1979. Sample Selection Bias as a Specification Error. *Econometrica* 47(1): 153–61.

Hendricks, Kenneth, and Robert H. Porter. 1996. The Timing and Incidence of Exploratory Drilling on Offshore Wildcat Tracts. *American Economic Review* 86(3): 388–407.

Hendry, David F. 1980. Econometrics: Alchemy or Science? *Economica* 47: 387–406.

Hendry, David F., and Mary S. Morgan, eds. 1995. *The Foundations of Econometric Analysis*. Cambridge: Cambridge University Press.

Heredity. 1947–62. Every full-length article published between 1947 and 1962.

Herrnstein, Richard J., and Charles Murray. 1994. *The Bell Curve*. New York: Free Press.

Hill, C. R., and B. Thompson. 2004. Computing and Interpreting Effect Size. Vol. 19, *Higher Education: Handbook of Theory and Research*, ed. J. C. Smart, 175–96. New York: Kluwer.

Hilts, Victor L. 1973. Statistics and Social Science. In Giere and Westfall, *Foundations of Scientific Method*, 206–33.

Hirshleifer, Jack. 2004. Personal communication, University of California, Los Angeles, January 5.

Hoel, Paul G. 1954. *Introduction to Mathematical Statistics*. New York: Wiley.

Hogg, Robert V., and Allen T. Craig. 1965. *Introduction to Mathematical Statistics*. 2nd ed. New York: Macmillan.

Hoover, Kevin, and Mark Siegler. 2008. Sound and Fury. Forthcoming. *Journal of Economic Methodology* 15 (March).

Horowitz, Joel L. 2004. Comments on "Size Matters." *Journal of Socio-Economics* 33(5): 551–54.

Horowitz, Joel L., M. Markatou, and R. V. Lenth. 1995. Robust Scale Estimation Based on the Empirical Characteristic Function. *Statistics and Probability Letters* 25:185–92.

Horowitz, Joel L., and V. G. Spokoiny. 2001. An Adaptive, Rate-Optimal Test of a Parametric Mean-Regression Model against a Nonparametric Alternative. *Econometrica* 69:599–631.

Hotelling, Harold. 1927. Review of Fisher's *Statistical Methods for Research Workers*. *Journal of American Statistical Association* 22:411–12.

Hotelling, Harold. 1928. Review of Fisher's *Statistical Methods for Research Workers*. *Journal of American Statistical Association* 23:346.

Hotelling, Harold. 1930a. Review of Fisher's *Statistical Methods for Research Workers*. *Journal of American Statistical Association* 25:381–82.

Hotelling, Harold. 1930b. British Statistics and Statisticians Today, *Journal of the American Statistical Association* 25 (170): 186–90.

Hotelling, Harold. 1931. The Generalization of Student's Ratio. *Annals of Mathematical Statistics* 2:360–78.

Hotelling, Harold. 1933. Review of Fisher's *Statistical Methods for Research Workers*. *Journal of American Statistical Association* 28:374–75.

Hotelling, Harold. 1935. Review of Fisher's *Statistical Methods for Research Workers*. *Journal of American Statistical Association* 30:771–72—Design.

Hotelling, Harold. 1937a. Review of Fisher's *Statistical Methods for Research Workers*. *Journal of American Statistical Association* 32:218–19.

Hotelling, Harold. 1937b. Review of Fisher's *Statistical Methods for Research Workers*. *Journal of American Statistical Association* 32:580–82—Design.

Hotelling, Harold. 1938a. Review of Fisher and Yates's, *Statistical Tables for Biological, Agricultural, and Medical Science*. *Science* 88:596–97.

Hotelling, Harold. 1938b. William Sealy Gosset, 1876–1937. *Journal of the Royal Statistical Society* 101(1): 248–49.

Hotelling, Harold. 1939. Review of Fisher's *Statistical Methods for Research Workers*. *Journal of American Statistical Association* 34:423–24.

Hotelling, Harold. 1940 [1960]. The Teaching of Statistics. *Annals of Mathematical Statistics* 11:457–70. Reprinted in Olkin et al., *Contributions to Probability and Statistics*, 11–24.

Housman, A. E. 1992. The Application of Thought to Textual Criticism. *The Classical Papers of A. E. Housman*, ed. J. Diggle and F. R. D. Goodyear, 1058–69.

Howie, David. 2002. *Interpreting Probability: Controversies and Developments in the Early Twentieth Century*. Cambridge: Cambridge University Press.

Howson, Colin, and Peter Urbach. 2006. *Scientific Reasoning: The Bayesian Approach*. Chicago: Open Court.

Hubbard, R., and P. A. Ryan. 2000. The Historical Growth of Statistical Significance Testing in Psychology—and Its Future Prospects. *Educational and Psychological Measurement* 60:661–81.

Huberty, C. J. 1999. On Some History Regarding Statistical Testing. In *Advances in Social Science Methodology*, ed. B. Thompson, 5:1–23. Stamford, CT: JAI Press.

Huberty, C. J. 2002. A History of Effect Size Indices. *Educational and Psychological Measurement* 62:227–40.

James, William. 1896 [1948]. The Will to Believe. In *Essays in Pragmatism*, 88–109. New York: Hafner.

Jeffrey, Richard. 1992. *Probability and the Art of Judgment*. Cambridge: Cambridge University Press.

Jeffreys, Harold. 1919. On the Crucial Test of Einstein's Theory of Gravitation. *Monthly Notices of the Royal Astronomical Society* 80 (December): 138–54.

Jeffreys, Harold. 1931 [1973]. *Scientific Inference*. Cambridge: Cambridge University Press.

Jeffreys, Harold. 1939a [1967]. *Theory of Probability*. 3rd ed. London: Oxford University Press.

Jeffreys, Harold. 1939b. Random and Systematic Arrangements. *Biometrika* 31(1–2): 1–8.

Jeffreys, Harold. 1963. Review of L. J. Savage et al., *The Foundations of Statistical Inference*. *Technometrics* 5(3): 407–10.

Jevons, William Stanley. 1874 [1877]. *The Principles of Science: A Treatise on Logic and Scientific Method*. 2nd ed. London: Macmillan.

Johnson, Norman L., and Samuel Katz, eds. 1997. *Leading Personalities in Statistical Sciences from the Seventeenth Century to the Present*. New York: Wiley.

Johnston, John. 1963, 1972, 1984. *Econometric Methods*. 1st, 2nd, 3rd editions. New York: McGraw-Hill.

Kachelmeier, Steven J., and Mohamed Shehata. 1992. Examining Risk Preferences under High Monetary Incentives: Experimental Evidence from the People's Republic of China. *American Economic Review* 82(5): 1120–41.

Kaye, D. H. 2002. The Error of Equal Error Rates. *Law, Probability, and Risk* 1(1): 3–8.

Kelly, Harold H. 1967. Attribution Theory in Social Psychology. Vol. 15 of *Nebraska Symposium on Motivation*, ed. D. Levine. Lincoln: University of Nebraska Press.

Kendall, M. C. 1952. George Udny Yule, 1871–1951. *Journal of the Royal Statistical Society*, Ser. A, 115:156–61.

Kennedy, Peter. 1985. *A Guide to Econometrics*. 2nd ed. Cambridge: M.I.T. Press.

Keuzenkamp, Hugo A. 2000. *Probability, Econometrics, and Truth*. Cambridge: Cambridge University Press.

Keuzenkamp, Hugo A., and Jan Magnus. 1995. On Tests and Significance in Econometrics. *Journal of Econometrics* 67(1): 103–28.

Kevles, Daniel J. 1985 [1995]. *In the Name of Eugenics*. Cambridge: Harvard University Press.

Keynes, John Maynard. 1921. *Treatise on Probability.* London: Macmillan.

King, Martin Luther, Jr. 1963 [1981]. *Strength to Love.* Philadelphia: Fortress.

Kirk, R. E. 2003. The Importance of Effect Magnitude. In *Handbook of Research Methods in Experimental Psychology,* ed. S. F. Davis, 83–105. Oxford: Blackwell.

Klamer, Arjo, D. N. McCloskey, and R. M. Solow, eds. 1989. *The Consequences of Economic Rhetoric.* Cambridge: Cambridge University Press.

Klein, Hugh, Kirk W. Elifson, and Claire E. Sterk. 2003. Perceived Temptation to Use Drugs and Actual Drug Use among Women. *Journal of Drug Issues* 33:161–92.

Klein, Lawrence. 1985. *Economic Theory and Econometrics.* Ed. J. Marquez. London: Blackwell.

Kmenta, Jan. 1971. *Elements of Econometrics.* New York: Macmillan.

Koopmans, Tjalling. 1937. *Linear Regression Analysis of Economic Time Series.* Haarlem: Netherlands Economic Institute.

Kravis, Irving, and Richard Lipsey. 1978. Price Behavior in the Light of Balance-of-Payments Theories. *Journal of International Economics* 8 (May): 193–246.

Krugman, Paul. 1978. Purchasing Power Parity: Another Look at the Evidence. *Journal of International Economics* 8 (August): 397–407.

Kruskal, William H. 1968a. Tests of Statistical Significance. In *International Encyclopedia of the Social Sciences,* ed. David Sills, 14:238–50. New York: Macmillan.

Kruskal, William H. 1968b. Statistics: The Field. In *International Encyclopedia of the Social Sciences,* ed. David Sills, 15:206–24. New York: Macmillan.

Kruskal, William H. 1978. Formulas, Numbers, Words: Statistics in Prose. *American Scholar* 47(2): 223–29.

Kruskal, William H. 1980. The Significance of Fisher: A Review of Joan Fisher Box's *R. A. Fisher: The Life of a Scientist. Journal of the American Statistical Association* 75(372): 1019–30.

Kruskal, William H. 2002. Interview with the author, August 16, 2002, University of Chicago.

Kruskal, William H., and Judith M. Tanur. 1968 [1978]. *International Encyclopedia of Statistics.* Vols. 1–2. New York: Free Press.

Kuhn, Thomas S. 1962 [1970]. *The Structure of Scientific Revolutions.* 2nd ed. Chicago: University of Chicago Press.

Lakatos, Imre. 1976. *Proofs and Refutations: The Logic of Mathematical Discovery.* Cambridge: Cambridge University Press.

Lancaster, Tony. 2004. *An Introduction to Modern Bayesian Econometrics.* Oxford: Blackwell .

Lang, Janet M., Kenneth J. Rothman, and Cristina I. Cann. 1998. That Confounded *p*-Value. *Epidemiology* 9(1): 7–8.

Lange, Oskar. 1959 [1978]. *Introduction to Econometrics.* Warsaw and Oxford: Polish Scientific Publishers and Pergamon.

Lanham, Richard A. 1991. *A Handlist of Rhetorical Terms.* Los Angeles: University of California Press.

Laslett, Barbara. 1990. Unfeeling Knowledge: Emotion and Objectivity in the History of Sociology. *Sociological Forum* 5 (September): 413–33.

Latter, Oswald. 1901. The Egg of *Cuculus Canorus:* An Enquiry into the Dimensions of the Cuckoo's Egg and the Relation of the Variations to the Size of the Eggs of the Foster Parent, with Notes on Coloration, etc. *Biometrika* 1(2): 164–76.

Leamer, Edward E. 1978. *Specification Searches: Ad Hoc Inference with Nonexperimental Data.* New York: Wiley.

Leamer, Edward E. 1982. Sets of Posterior Means with Bounded Variance Priors. *Econometrica* (May): 725–36.

Leamer, Edward E. 2004. Are the Roads Red? Comments on "Size Matters." *Journal of Socio-Economics* 33(5): 555–58.

Leamer, Edward E., and Herman B. Leonard. 1983. Reporting the Fragility of Regression Estimates. *Review of Economics and Statistics* LXV (May): 306–17.

Lehmann, Erich L. 1959. *Testing Statistical Hypotheses.* New York: Wiley.

Lehmann, Erich L. 1968. Hypothesis Testing. In Kruskal and Tanur, *International Encyclopedia of Statistics,* vol. 1:441–49.

Lehmann, Erich L. 1993. The Fisher, Neyman-Pearson Theories of Testing Hypotheses: One Theory or Two? *Journal of the American Statistical Association* 88:1242–49.

Lehmann, Erich L. 1999. "Student" and Small-Sample Theory. *Statistical Science* 14(4): 418–26.

Levy, David M. 2004. *How the Dismal Science Got Its Name: Classical Economics and the Ur-Text of Racial Politics.* Ann Arbor: University of Michigan Press.

Lindley, Dennis V. 1990. Fisher: A Retrospective. *Chance* 3(1):31–32.

Lindley, Dennis V. 1991. Sir Harold Jeffreys. *Chance* 4(2): 10–14, 21.

Lisse, Jeffrey R., Monica Perlman, Gunnar Johansson, James R. Shoemaker, Joy Schectman, Carol S. Skalky, Mary E. Dixon, Adam B. Polis, Arthur J. Mollen, and Gregory P. Geba. 2003. Gastrointestinal Tolerability and Effectiveness of Rofecoxib (Vioxx®) versus Naproxen in the Treatment of Osteoarthritis. *Annals of Internal Medicine* 139 (October): 539–46.

Loftus, Geoffrey R. 1993. Editorial comment. *Memory and Cognition* 21:1–3.

Lowe, G. R., and P. N. Reid, eds. 1999. *The Professionalization of Poverty.* New York: Aldine de Gruyter.

Luby, S. P., J. L. Jones, and J. M. Horan. 1993. A Large Salmonellosis Outbreak Associated with a Frequently Penalized Restaurant. *Epidemiology and Infection* 110:31–39.

Lucas, Robert E., and Thomas J. Sargent. 1981. *Rational Expectations and Econometric Practice.* Minneapolis: University of Minnesota Press.

Lunt, Peter. 2004. The Significance of the Significance Test Controversy: Comments on "Size Matters." *Journal of Socio-Economics* 33(5): 559–64.

Lupton, Sydney. 1898. *Notes on Observations.* London: MacMillan.

Lynch, Patrick, and John Vaizey. 1960. *Guinness Brewery in the Irish Economy, 1759–1876.* Cambridge: Cambridge University Press.

MacDonnell, William R. 1901. On Criminal Anthropometry and the Identification of Criminals. *Biometrika* 1(2): 177–227.

MacKenzie, Donald A. 1981. *Statistics in Britain, 1865–1930: The Social Construction of Scientific Knowledge.* Edinburgh: Edinburgh University Press.

Maddala, G. S., C. R. Rao, and H. D. Vinod. 1993. *Handbook of Statistics*. Vol. 11: *Econometrics*. Amsterdam: North-Holland.

Mahalanobis, P. C. 1938. Professor Ronald Aylmer Fisher. In R. A. Fisher, *Contributions to Mathematical Statistics,* ed. Walter A. Shewhart, 365–72. New York: Wiley & Sons, 1950.

Mahalanobis, P. C. 1964. In Memoriam: Ronald Aylmer Fisher, 1890–1962. *Biometrics* 20(2): 368–71.

Marshall, Stephen W. 2005. Commentary on [Batterham and Hopkins]. *Sport-Science* 9 (December): 43–44.

Mathias, Peter. 1979. *The Transformation of England*. New York: Columbia University Press.

Matthews, R. 1998. The Great Health Hoax. *Sunday Telegraph,* September 13.

Mayer, Thomas. 1979. *Economics as an Exact Science: Realistic Goal or Wishful Thinking?* Working Papers, no. 124. Davis: Department of Economics, University of California, Davis.

Mayo, Deborah G. 1999. *Error and the Growth of Experimental Knowledge*. Chicago: University of Chicago Press.

Mayr, Ernst. 1982. *The Growth of Biological Thought: Diversity, Evolution, and Inheritance*. Cambridge: Belknap and Harvard University Press.

McBurney, Peter, and Simon Parsons. 2002. Determining Error Bounds for Hypothesis Tests in Risk Assessment: A Research Agenda. *Law, Probability, and Risk* 1(1): 17–36.

McCarthy, Michael A., and Kirsten M. Parris. 2004. Clarifying the Effect of Toe Clipping on Frogs with Bayesian Statistics. *Journal of Applied Ecology* 41:780–86.

McCloskey, Deirdre, and J. Richard Zecher. 1976. How the Gold Standard Worked, 1880–1913. In *The Monetary Approach to the Balance of Payments,* ed. J. A. Frenkel and H. G. Johnson, 192–208. London: Allen and Unwin.

McCloskey, Deirdre. 1985a. The Loss Function Has Been Mislaid: The Rhetoric of Significance Tests. *American Economic Review,* supplement, 75(2): 201–5.

McCloskey, Deirdre. 1985b [1998]. *The Rhetoric of Economics*. Madison: University of Wisconsin Press. Especially chapters 7 and 8 of the 1985 edition and chapters 8 and 9 of the 1998 edition.

McCloskey, Deirdre. 1992. The Bankruptcy of Statistical Significance. *Eastern Economic Journal* 18 (summer): 359–61.

McCloskey, Deirdre. 1995. The Insignificance of Statistical Significance. *Scientific American,* April, 32–33.

McCloskey, Deirdre. 1996. Why Economic Historians Should Stop Relying on Statistical Tests of Significance and Lead Economists and Historians into the Promised Land. *Newsletter of the Cliometric Society* 2(2): 5–7.

McCloskey, Deirdre. 2002. The Insanity of Letters of Recommendation. *Chronicle of Higher Education,* January.

McCloskey, Deirdre. 2006. *Bourgeois Virtues*. Chicago: University of Chicago Press.

McCloskey, Deirdre, and Stephen Ziliak. 2008. Signifying Nothing: Reply to Hoover and Seigler. Forthcoming, *Journal of Economic Methodology* 15 (March).

McCloskey, Deirdre, and Stephen T. Ziliak. 1996. The Standard Error of Regressions. *Journal of Economic Literature* 34 (March): 97–114.

McMullen, Launce. 1939. "Student" as a Man. *Biometrika* 30(3–4): 205–10.

Meehl, Paul E. 1954. *Clinical versus Statistical Prediction: A Theoretical Analysis and Review of the Evidence*. Minneapolis: University of Minnesota Press.

Meehl, Paul E. 1967. Theory-Testing in Psychology and Physics: A Methodological Paradox. *Philosophy of Science* 34:103–115. Reprinted in Morrison and Henkel, *The Significance Test Controversy*, 252–66.

Meehl, Paul E. 1978. Theoretical Risks and Tabular Asterisks: Sir Karl, Sir Ronald, and the Slow Progress of Soft Psychology. *Journal of Consulting and Clinical Psychology* 46:806–34.

Meehl, Paul E. 1990. Why Summaries of Research on Psychological Theories Are Often Uninterpretable. *Psychological Reports* 66(144): 195–244.

Meehl, Paul E. 1998. The Power of Quantitative Thinking. Speech delivered upon receipt of the James McKeen Cattell Fellow Award at the annual meetings of the American Psychological Society, Washington, DC, May 23.

Melton, Arthur W. 1962. Editorial. *Journal of Experimental Psychology* 64(2): 553–57.

Mendel, Gregor. 1955. *Experiments in Plant Hybridisation*. Edinburgh: Oliver and Boyd.

Merton, Robert M. 1949 [1957]. *Social Theory and Social Structure*. New York: Free Press.

Mirowski, Phillip, and D. Wade Hands. 1998. A Paradox of Budgets: The Postwar Stabilization of American Neoclassical Demand Theory. In *From Interwar Pluralism to Postwar Neoclassicism*, ed. M. S. Morgan and M. Rutherford, 260–92. Durham: Duke University Press. Annual supplement to *History of Political Economy*.

Mohr, Lawrence B. 1990. *Understanding Statistical Significance*. Quantitative Applications in the Social Sciences. Thousand Oaks, CA: Sage.

Mood, Alexander M. 1950. *Introduction to the Theory of Statistics*. New York: McGraw-Hill.

Mood, Alexander M., and F. A. Graybill. 1963. *Introduction to the Theory of Statistics*. 2nd ed. New York: McGraw-Hill.

Moore, David S., and George P. McCabe. 1993. *Introduction to the Practice of Statistics*. New York: Freeman.

Moran, J. L., and P. J. Soloman. 2004. A Farewell to *p*-Values? *Critical Care and Resuscitation* 6:130–37.

Morgan, Mary S. 1990. *The History of Econometric Ideas*. Cambridge: Cambridge University Press.

Morgenstern, Oskar von. 1963. *On the Accuracy of Economic Observations*. 2nd ed. Princeton: Princeton University Press.

Morrison, Denton E., and Ramon E. Henkel. 1969. Significance Tests Reconsidered. *American Sociologist* 4 (May): 131–40.

Morrison, Denton E., and Ramon E. Henkel. 1970. *The Significance Test Controversy: A Reader*. Chicago: Aldine.

Mosteller, Frederick. 1964. Samuel S. Wilks: Statesman of Statistics. *American Statistician* 18 (2): 11–17.

Mosteller, Frederick. 1968. Nonsampling Errors. In Kruskal and Tanur, *International Encyclopedia of Statistics,* 2:208–29.

Mosteller, Frederick. 2005. Personal communication, memorial service for William H. Kruskal, University of Chicago, May 21, 2005.

Mosteller, Frederick, and R. R. Bush. 1954. Selected Quantitative Techniques. In *Handbook of Social Psychology,* ed. Gardner Lindzey, 289–334. Cambridge, MA: Addison-Wesley.

Murdoch, Iris. 1967. The Sovereignty of Good over Other Concepts. 77–104 in Murdoch, *The Sovereignty of Good.* London: Routledge.

New England Journal of Medicine. 1970–2005. Every "original article" published between August 1970 and August 2005 that uses quantitative methods.

Neyman, Jerzy. 1938. L'estimation statistique traite comme un probleme classique de probabilité. *Actualites Scientifiques et Industrielles* 739:25–57.

Neyman, Jerzy. 1956. Note on an Article by Sir Ronald Fisher. *Journal of the Royal Statistical Society.* Ser. B (Methodological) 18(2): 288–94.

Neyman, Jerzy. 1957. "Inductive Behavior" as a Basic Concept of Philosophy of Science. *Review of the Mathematical Statistics Institute* 25:7–22.

Neyman, Jerzy. 1960. Harold Hotelling: A Leader in Mathematical Statistics. In Olkin et al., *Contributions to Probability and Statistics,* 6–10.

Neyman, Jerzy. 1961. Silver Jubilee of My Dispute with Fisher. *Journal of Operations Research* (Japan) 3:145–54.

Neyman, Jerzy. 1981. Egon S. Pearson (August 11, 1895–June 12, 1980): An Appreciation. *Annals of Statistics* 9(1): 1–2.

Neyman, Jerzy, and E. S. Pearson. 1928. On the Use and Interpretation of Certain Test Criteria for Purposes of Statistical Inference, Part I and Part II. *Biometrika* 20:175–294.

Neyman, Jerzy, and E. S. Pearson. 1933. On the Problem of the Most Efficient Tests of Statistical Hypotheses. *Phil. Trans. Roy. Soc.* Ser. A, 231:289–337.

Neyman, Jerzy, and E. S. Pearson. 1938. Note on Some Points on "Student's" Paper on "Comparison between Balanced and Random Arrangements of Field Plots." *Biometrika* 29(3–4): 379–88.

O'Brien, Anthony P. 2004. Why Is the Standard Error of Regression So Low Using Historical Data? *Journal of Socio-Economics* 33(5): 565–70.

Olkin, Ingram, S. G. Ghurye, W. Hoeffding, W. G. Madow, and H. B. Mann, eds. 1960. *Contributions to Probability and Statistics: Essays in Honor of Harold Hotelling.* Stanford: Stanford University Press.

Orton, Clive. 1997. Testing Significance or Testing Credulity? *Oxford Journal of Archaeology* 16 (July): 219.

Osterweil, Neil. 2006. Atypical Antipsychotics Get Poor Grade for Alzheimer's Psychoses. *Psychiatric Times,* October 11. http://psychiatrictimes.com.

Parkhurst, David F. 1997. Disadvantages of Significance Testing. http://www.indiana.edu/~stigtsts.

Pearson, Egon S. 1936. Karl Pearson: An Appreciation of His Life and Work, Part 1: 1857–1906. *Biometrika* 28(3–4): 193–257.

Pearson, Egon S. 1938. Karl Pearson: An Appreciation of His Life and Work, Part 2: 1906–1936. *Biometrika* 29(3–4): 161–248.

Pearson, Egon S. 1939. "Student" as Statistician. *Biometrika* 30 (3–4): 210–50.

Pearson, Egon S. 1966. The Neyman-Pearson Story, 1926–34. In David, *Research Papers in Statistics*, 1–24.

Pearson, Egon S. 1968. Studies in the History of Probability and Statistics, Part 20: Some Early Correspondence between W. S. Gosset, R. A. Fisher, and Karl Pearson with Notes and Comments. *Biometrika* 55(3): 445–57.

Pearson, Egon S. 1990 [posthumous]. *"Student": A Statistical Biography of William Sealy Gosset*. Edited and augmented by R. L. Plackett with the assistance of G. A. Barnard. Oxford: Clarendon.

Pearson, Egon S., and N. K. Adyanthaya. 1929. The Distribution of Frequency Constants in Small Samples from Non-normal Symmetrical and Skew Populations. *Biometrika* 21(1–4): 259–86.

Pearson, Egon S., and M. C. Kendall, eds. 1970. *Studies in the History of Statistics and Probability*. London: Charles Griffin.

Pearson, Karl. 1892 [1951]. *The Grammar of Science*. London: Arrowsmith.

Pearson, Karl. 1914. *Tables for Biometricians and Statisticians*. London: Biometrics Laboratory, University College.

Pearson, Karl. 1931. Further Remarks on the "*z*" Test. *Biometrika* 23(3–4): 408–15.

Pearson, Karl, Francis Galton, William Weldon, and Charles Davenport. 1901. Editorial: The Scope of Biometrika. *Biometrika* 1(1): 1–2.

Pearson, Karl, and Alice Lee. 1903. On the Laws of Inheritance in Man, Part 1: Inheritance of Physical Characters. *Biometrika* 2(4): 357–462.

Pearson, Karl, and Margaret Moul. 1925. The Problem of Alien Immigration into Great Britain, Illustrated by an Examination of Russian and Polish Jewish Children. *Annals of Eugenics* 1(1–2): 5–127.

Pearson, Karl, and Margaret Moul. 1927. The Mathematics of Intelligence: The Sampling Errors in the Theory of a Generalized Factor. *Biometrika* 19(3–4): 246–91.

Peart, Sandra J., and David M. Levy. 2005. *The "Vanity of the Philosopher": From Equality to Hierarchy in Post-classical Economics*. Ann Arbor: University of Michigan Press.

Peterman, R. M. 1990. Statistical Power Analysis Can Improve Fisheries Research and Management. *Canadian Journal of Fisheries and Aquatics Management* 47: 2–15.

Polanyi, Michael. 1959. *The Study of Man*. Chicago: University of Chicago Press.

Polley, William J. 2003. Measurement Ahead of Theory? The Economic Significance of Recent Research in Time Series Modeling. www.williampolley.com.

Popper, Karl R. 1934 [1959]. *The Logic of Scientific Discovery*. London and New York: Routledge.

Porter, Theodore M. 1986. *The Rise of Statistical Thinking, 1820–1900*. Princeton: Princeton University Press.

Porter, Theodore M. 1995. *Trust in Numbers: The Pursuit of Objectivity in Science and Public Life*. Princeton: Princeton University Press.

Porter, Theodore M. 2004. *Karl Pearson: The Scientific Life in a Statistical Age*. Princeton: Princeton University Press.

Pratt, J. W., H. Raiffa, and R. Schlaifer. 1995. *Introduction to Statistical Decision Theory*. Cambridge: MIT Press.

Prener, Anne, G. Enghom, and O. M. Jensen. 1996. Genital Anomalies and Risk for Testicular Cancer in Danish Men. *Epidemiology* 7(1): 14–19.

Press, S. James. 1972 [2005]. *Applied Multivariate Analysis: Using Bayesian and Frequentist Methods of Inference.* 2nd ed. Mineola, NY: Dover.

Press, S. James. 2003. *Subjective and Objective Bayesian Statistics.* New York: Wiley.

Quine, W. V., and J. S. Ullian. 1970 [1978]. *The Web of Belief.* New York: Mc-Graw-Hill.

Raiffa, Howard, and Robert Schlaifer. 1961. *Applied Statistical Decision Theory.* New York: HarperCollins.

Reid, Constance. 1982 [1998]. *Neyman: A Life.* New York: Springer.

Rennie, Drummond. 1978. Vive la Difference ($p < 0.05$). *New England Journal of Medicine* 299:828–29.

Richardson, J. D. 1978. Some Empirical Evidence on Commodity Arbitrage and the Law of One Price. *Journal of International Economics* 8 (May): 341–51.

Ridker, Paul M., et al. 2005. A Randomized Trial of Low-Dose Aspirin in the Primary Prevention of Cardiovascular Disease in Women. *New England Journal of Medicine* 352 (13): 1293–1304.

Robinson, D. H., and J. Levin. 1997. Reflections on Statistical and Substantive Significance with a Slice of Replication. *Educational Researcher* 26(5): 21–26.

Robinson, D. H., and H. Wainer. 2002. On the Past and Future of Null Hypothesis Significance Testing. *Journal of Wildlife Management* 66:263–71.

Romer, Cristina D. 1986. Is the Stabilization of the Postwar Economy a Figment of the Data? *American Economic Review* 76 (3): 314–34.

Rosen, Steven A. 1986. A Note on Frequencies, Proportions, and Diversity: A Response to Cannon. *American Antiquity* 51(2): 409–11.

Rosenthal, R. 1994. Parametric Measures of Effect Size. In *The Handbook of Research Synthesis,* ed. H. Cooper and L. V. Hedges, 231–44. New York: Russell Sage Foundation.

Ross, Dorothy. 1991. *The Origins of American Social Science.* Cambridge: Cambridge University Press.

Rossi, Joseph. 1990. Statistical Power of Psychological Research: What Have We Gained in 20 Years? *Journal of Consulting and Clinical Psychology* 58:646–56.

Rothman, Kenneth J. 1978. A Show of Confidence. *New England Journal of Medicine* 299:1362–63.

Rothman, Kenneth J. 1986. *Modern Epidemiology.* New York: Little, Brown.

Rothman, Kenneth J. 1998. Writing for Epidemiology. *Epidemiology* 9(3): 333–37.

Rothman, Kenneth J. 2002. *Epidemiology.* New York: Oxford University Press.

Rothman, Kenneth J., Eric S. Johnson, and David S. Sugano. 1999. Is Flutamide Effective in Patients with Bilateral Orchiectomy? *Lancet* 353:1184.

Rozeboom, William W. 1960. The Fallacy of the Null-Hypothesis Significance Test. *Psychological Bulletin* 57:416–28.

Rozeboom, William W. 1997. Good Science Is Abductive Not Hypothetico-Deductive. In Harlow, Mulaik, and Staiger, *What If There Were No Significance Tests?* 335–92.

Sachs, Jeffrey. 1980. The Changing Cyclical Behavior of Wages and Prices: 1880–1976. *American Economic Review* 70(1): 78–90.

Sargent, Thomas J. 1981. Rational Expectations, the Real Rate of Interest, and the Natural Rate of Unemployment. In Lucas and Sargent, *Rational Expectations and Econometric Practice*, 159–98.

Sartre, Jean-Paul. 1948 [1995]. *Anti-Semite and Jew*. New York: Schocken.

Savage, Leonard J. 1954 [1972]. *The Foundations of Statistics*. New York: Dover.

Savage, Leonard J. 1971a [1976, posthumous]. On Re-reading R. A. Fisher. *Annals of Statistics* 4(3): 441–500.

Savage, Leonard J. 1971b [1974, posthumous]. Elicitation of Personal Probabilities. In Fienberg and Zellner, *Studies in Bayesian Econometrics in Honor of Leonard J. Savage*, 111–56.

Savin, N. E., and A. E. Würtz. 1999. Power of Tests in Binary Response Models. *Econometrica* 67:413–22.

Savitz, David A., K. Tolo, and C. Poole. 1990. Statistical Significance Testing in the *American Journal of Epidemiology* 139:1047–52.

Schelling, Thomas. 2004. Correspondence. *Econ Journal Watch* 1(3): 539–40. http://www.econjournalwatch.com. Comment on Ziliak and McCloskey, "Size Matters."

Schlaifer, Robert. 1959. *Probability and Statistics for Business Decisions*. New York: McGraw-Hill.

Schmidt, Frank L. 1996. Statistical Significance Testing and Cumulative Knowledge in Psychology: Implications for the Training of Researchers. *Psychological Methods* 1:115–29.

Schmidt, Frank L., and J. E. Hunter. 1997. Eight Common but False Objections to the Discontinuation of Significance Testing in the Analysis of Research Data. In Harlow, Mulaik, and Steiger, *What If There Were No Significance Tests?* 37–64.

Schwartz, Joel. 1996. Air Pollution and Hospital Admissions for Respiratory Disease. *Epidemiology* 7:20–28.

Scott, Elizabeth. 1953. Testing Hypotheses. In *Statistical Astronomy*, ed. Robert J. Trumpler and Harold F. Weaver, 220–30. New York: Dover.

Sedlmeier, P., and G. Gigerenzer. 1989. Do Studies of Statistical Power Have an Effect on the Power of Studies? *Psychological Bulletin* 105:309–15.

Shelton, R. C., M. B. Keller, A. Gelenberg, et al. 2001. Effectiveness of St. John's Wort in Major Depression: A Randomized Controlled Trial, *Journal of the American Medical Association* 285:178–86.

Shewhart, Walter A. 1929. Basis for Economic Control of Quality. Bell Telephone Laboratories, Inspection Engineering Department.

Shewhart, Walter A. 1931. *Economic Control of Quality of Manufactured Product*. New York: Van Nostrand.

Shiller, Robert. 1987. The Volatility of Stock Market Prices. *Science* 235:33–36.

Shulman, L. S. 1970. Reconstruction of Educational Research. *Review of Educational Research* 40:371–93.

Shyrock, Richard H. 1961. The History of Quantification in Medical Science. *Isis* 52(2): 215–37.

Simonoff, Jeffrey. 2003. *Analyzing Categorical Data*. New York: Springer-Verlag.

Smith, Adam. 1759 [1982]. *The Theory of Moral Sentiments*. Ed. D. D. Raphael and A. L. Macfie. Glasgow ed. Indianapolis: Liberty Classics, 1982.

Snowdon, B., and H. Vane. 1999. *Conversations with Leading Economists.* Cheltenham: Edward Elgar.

Solon, Gary. 1992. Intergenerational Income Mobility in the United States. *American Economic Review* 82(3): 393–408.

Solow, Robert. 1988 [1997]. Robert Solow. In *Lives of the Laureates,* ed. W. Breit and R. W. Spencer, 183–204. Cambridge: MIT Press.

Spanos, Aris. 1986. *Statistical Foundations of Econometric Modeling.* Cambridge: Cambridge University Press.

Spanos, Aris. 1999. *Probability Theory and Statistical Inference.* Cambridge: Cambridge University Press.

Sterling, Thomas D. 1959. Publication Decisions and Their Possible Effects on Inferences Drawn from Tests of Significance—or Vice Versa. *Journal of the American Statistical Association* 54(285): 30–34.

Sterne, J. A. C., and G. Davey Smith. 2001. Sifting the Evidence: What's Wrong with Significance Tests? *British Medical Journal* 322:226–31.

Steward, Dwight, and S. W. O'Donnell. 2005. Evaluating the Statistical and Economic Significance of Statistical Evidence in Employment Discrimination Cases. *Expert Evidence Report* 5(5): 117–. http://www.bna.com/.

Stigler, Stephen. 1986. *The History of Statistics.* Cambridge: Harvard University Press.

Stigler, Stephen. 1996. The History of Statistics in 1933. *Statistical Science* 11:244–52.

Stigler, Stephen. 1999. *Statistics on the Table.* London: Harvard University Press.

Stigler, Stephen. 2000. The Problematic Unity of Biometrics. *Biometrics* 56(3): 272–77.

Stigler, Stephen. 2005a. Fisher in 1921. *Statistical Science* 20(1): 32–49.

Stigler, Stephen. 2005b. Discussion of D. Denis. *Journal de la Societé Francaise de Statistique* 145(4): 63–64.

Student [W. S. Gosset]. 1904. The Application of the "Law of Error" to the Work of the Brewery. Report, Arthur Guinness Son and Co., 3 November. In E. Pearson, "Student" as Statistician, 212–15.

Student. 1905. The Pearson Co-efficient of Correlation. Supplement to Report of 1904, Arthur Guinness Son and Co. In E. Pearson, "Student" as Statistician, 217.

Student. 1907. On the Error of Counting with a Haemacytometer. *Biometrika* 5(3): 351–60.

Student. 1908a. The Probable Error of a Mean. *Biometrika* 6(1): 1–24.

Student. 1908b. The Probable Error of a Correlation Coefficient. *Biometrika* 6(2/3): 300–310.

Student. 1917. Tables for Estimating the Probability That the Mean of a Unique Sample of Observations Lies between $-\infty$ and Any Given Distance of the Mean of the Population from Which the Sample Is Drawn. *Biometrika* 11(4): 414–17.

Student. 1923. On Testing Varieties of Cereals. *Biometrika* 15 (3–4): 271–93.

Student. 1925. New Tables for Testing the Significance of Observations. *Metron* 5(3): 105–8.

Student. 1926 [1942]. Mathematics and Agronomy. *Journal of the American*

Society of Agronomy 18. Reprinted in Student, *Student's Collected Papers*, 121–34.

Student. 1927. Errors of Routine Analysis. *Biometrika* 19(1–2): 151–64.

Student. 1931a. The Lanarkshire Milk Experiment. *Biometrika* 23(3–4): 398–406.

Student. 1931b. On the "*z*" Test. *Biometrika* 23(3–4): 407–8.

Student. 1936. Co-operation in Large-Scale Experiments. *Supplement to Journal of the Royal Statistical Society* 3(2): 115–36.

Student. 1938 [posthumous]. Comparison between Balanced and Random Arrangements of Field Plots. *Biometrika* 29 (3–4): 363–78.

Student. 1942 [posthumous]. Student's *Collected Papers*. Ed. E. S. Pearson and John Wishart. London: Biometrika Office.

Taylor, B. L., and T. Gerrodette. 2004. The Uses of Statistical Power in Conservation Biology: The Vaquita and Northern Spotted Owl. *Conservation Biology* 7:489–500.

Thompson, Bruce. 1996. AERA Editorial Policies regarding Statistical Significance Testing: Three Suggested Reforms. *Educational Researcher* 25(2): 26–30.

Thompson, Bruce. 1997. Editorial Policies regarding Statistical Significance Tests: Further Comments. *Educational Researcher* 26(5): 29–32.

Thompson, Bruce. 2002a. Statistical, Practical, and Clinical: How Many Kinds of Significance Do Counselors Need to Consider? *Journal of Counseling and Development* 80:64–71.

Thompson, Bruce. 2002b. What Future Quantitative Social Science Research Could Look Like: Confidence Intervals for Effect Sizes. *Educational Researcher* 31(3): 24–31.

Thompson, Bruce. 2004. The "Significance" Crisis in Psychology and Education. *Journal of Socio-Economics* 33(5): 607–13.

Thompson, Bruce. In press. Research Synthesis: Effect Sizes. In *Complementary Methods for Research in Education*, ed. J. Green, G. Camilli, and P. B. Elmore. Washington, DC: American Educational Research Association.

Thompson, Bruce. In press. Effect Sizes vs. Statistical Significance. In *Research in Organizations: Foundational Principles, Processes, and Methods of Inquiry*, ed. R. A. Swanson and E. F. Holton. San Francisco: Berrett-Koehler.

Thompson, William L. 2005. 326 Articles/Books Questioning the Indiscriminate Use of Statistical Hypothesis Tests in Observational Studies. http:// www.cnr .colostate.edu/ ~anderson/ thompson1.html

Thorbecke, Erik. 2004. Economic and Statistical Significance: Comments on "Size Matters." *Journal of Socio-Economics* 33(5): 571–76.

Tippett, Leonard H. C. 1952. *The Methods of Statistics*. 4th ed. New York: Wiley.

Tukey, John W. 1969. Analyzing Data: Sanctification or Detective Work? *American Psychologist* 24:83–91.

Tullock, Gordon. 1959. Publication Decisions and Tests of Significance: A Comment. *Journal of the American Statistical Association* 54(287): 593.

Tversky, A., and D. Kahneman. 1971. Belief in the Law of Small Numbers. *Psychological Bulletin* 76:105–10.

Twain, Mark. 1883 [1946]. *Life on the Mississippi*. New York: Bantam.

Tyler, R. W. 1931. What Is Statistical Significance? *Educational Research Bulletin* 10:115–18, 142.

Vacha-Haase, T., J. E. Nilsson, D. R. Reetz, T. S. Lance, and B. Thompson. 2000. Reporting Practices and APA Editorial Policies Regarding Statistical Significance and Effect Size. *Theory and Psychology* 10: 413–25.

Vacha-Haase, T., and Bruce Thompson. 2004. How to Estimate and Interpret Various Effect Sizes. *Journal of Counseling Psychology* 51:473–81.

Veblen, Thorstein. 1914. *The Instinct of Workmanship*. New York: Macmillan.

Vining, Rutledge. 1949. Koopmans and the Choice of Variables to Be Studied and of Methods of Measurement. *Review of Economics and Statistics* 31 (May): 77–86.

Wald, Abraham. 1939. Contributions to the Theory of Statistical Estimation and Testing Hypotheses. *Annals of Mathematical Statistics* 10(4): 299–326.

Wald, Abraham. 1950. *Statistical Decision Functions*. New York: Wiley.

Wallis, W. Allen, and Harry V. Roberts. 1956. *Statistics: A New Approach*. Glencoe: Free Press.

Walster, G. William, and T. Anne Cleary. 1970. A Proposal for a New Editorial Policy in the Social Sciences. *American Statistician* 24(2): 16–19.

White, James Boyd. 1989. *Justice as Translation*. Chicago: University of Chicago Press.

Wilkonson, L., and APA Task Force on Statistical Inference. 1999. Statistical Methods in Psychology Journals. *American Psychologist* 54: 594–604.

Wilks, Samuel S. 1948 [1966]. *Elementary Statistical Analysis*. Princeton: Princeton University Press.

Wishart, John. 1938. William Sealy Gosset, 1876–1937. *Journal of the Royal Statistical Society* 101(1): 249–51.

Wonnacott, Ronald J., and Thomas H. Wonnacott. 1982. *Statistics: Discovering Its Power*. New York: Wiley.

Woofter, T. J., Jr. 1933. Common Errors in Sampling. *Social Forces* 11(4): 521–25.

Wooldridge, Jeffrey M. 2000. *Introductory Econometrics*. South-Western College Publishing.

Wooldridge, Jeffrey M. 2004. Statistical Significance Is Okay, Too: Comment on "Size Matters." *Journal of Socio-Economics* 33(5): 577–80.

Wrinch, Dorothy, and Harold Jeffreys. 1921. The Relationship between Geometry and Einstein's Theory of Gravitation. *Nature* 106:806–9.

Würtz, A. H. 2003. A Universal Upper Bound on Power Functions. UNSW Discussion Papers, 97/17. http://www.econ.ku.dk/wurtz/#Research.

Xanax advertisement. 1983. *Journal of Clinical Psychiatry* 44(7): 255–56.

Yates, Frank. 1951. The Influence of *Statistical Methods for Research Workers* on the Development of the Science of Statistics. *Journal of the American Statistical Association* 46:19–34.

Yates, Frank, and Kenneth Mather. 1963. Ronald Aylmer Fisher, *Biograph. Mem. Fell. R. Soc. Lond.* 9:91–129.

Yule, G. Udny, and M. G. Kendall. 1937. *An Introuduction to the Theory of Statistics*. London: Charles Griffin.

Zabell, Sandy. 1989. R. A. Fisher on the History of Inverse Probability. *Statistical Science* 4(3): 247–63.

Zellner, Arnold. 1984. *Basic Issues in Econometrics*. Chicago: University of Chicago Press.

Zellner, Arnold. 1997. *Bayesian Analysis in Econometrics and Statistics.* Cheltenham, UK: Elgar.

Zellner, Arnold. 2004a. To Test or Not to Test and If So, How? Comments on "Size Matters." *Journal of Socio-Economics* 33(5): 581–86.

Zellner, Arnold. 2004b. Personal communication, University of Chicago, May 26.

Zellner, Arnold. 2005a. Some Thoughts about S. James Press and Bayesian Analysis. Paper prepared for presentation at the Conference in Honor of Professor S. James Press, University of California, Riverside, May 14.

Zellner, Arnold. 2005b. Personal communication, University of Chicago, October 20.

Zellner, Arnold. 2008. Generalizing the Standard Product Rule of Probability Theory and Bayes's Theorem. Forthcoming, *Journal of Econometrics.*

Zellner, Arnold, ed. 1980. *Bayesian Analysis in Econometrics and Statistics: Essays in Honor of Harold Jeffreys.* Amsterdam: North-Holland.

Ziliak, Stephen T. 2002. Pauper Fiction in Economic Science: Paupers in Almshouses and the Odd Fit of *Oliver Twist. Review of Social Economy* 60(2): 159–81.

Ziliak, Stephen T. 2004. The Significance of the Economics Research Paper. In *A Guide to What's Wrong with Economics,* ed. Edward Fullbrook, 223–36. London: Anthem Press.

Ziliak, Stephen T. 2005. Why I Left Alan Greenspan to Seek *Economic* Significance: The Confessions of an α-Male. *Rethinking Marxism* 17 (January): 45–58.

Ziliak, Stephen T. 2008a. Guinessometrics: The Economic Foundation of 'Student's' t. *Journal of Economic Perspectives.* Forthcoming.

Ziliak, Stephen T. 2008b. "Student's" Methods: Science Before Fisher. Book manuscript, Roosevelt University, Department of Economics.

Ziliak, Stephen T., and Deirdre N. McCloskey. 2004a. Size Matters: The Standard Error of Regressions in the *American Economic Review. Journal of Socio-Economics* 33(5): 527–46.

Ziliak, Stephen T., and Deirdre N. McCloskey. 2004b. Significance Redux. *Journal of Socio-Economics* 33(5): 665–75.

Ziliak, Stephen T. (with Joan Hannon). 2006. Public Assistance: Colonial Times to the 1920s. Vol. 2, Chp. Bf, in *Historical Statistics of the United States,* ed., Susan Carter et al. New York: Cambridge University Press.

Zimmerman, David J. 1992. Regression toward Mediocrity in Economic Stature. *American Economic Review* 82(3): 409–29.

Index

abduction, the standard error compared to pragmatic, 152. *See also* beta

abnormal psychology, schizoid use of significance testing in, 135–38, 156

Adams, Henry, admires Karl Pearson's *Grammar of Science,* 196

addition rules: abstract numbers added differently by Fisher and Fisherians, 10; identical operation of if counting money or abstract numbers, 10

Administrative Science Quarterly, management journal, into the crisis, 117, 118, 248

Adyanthaya, N. K., and Egon S. Pearson, first power function estimated in 1929 by, 69

Agee, James, poet, specification error as betrayal of subject, 57, 60

Airy, George B., Royal Astronomer, inventor of clock at Prime Meridian, textbook teaches least squares method and theory of errors of observations to Gosset, 84, 207; uses modulus, 188

Allen, Robert, exemplary article in economics, 90

alpha, alpha-statistician (α-self, α-Fisher, wasp): α-only lacks or misapplies judgment, Savage, 15, 59, 60, 134, 180; does not understand opportunity cost, Gosset

offsets Fisher's with β-statistician, 60; as irrational skeptic, 197; Fisher invents in angry letter to β-Gosset, 59, 60, 217; Fisher institutionalizes in the sciences, 141, 216, 217, 222–26; Fisher only makes time for his, 214, 218; Hotelling adopts and teaches to leading statisticians, economists, and political scientists, 233; Karl Pearson adopts and teaches to Gosset's class, 197; as sizeless scientist, 5–21. *See also* Statistics of Unprogressive Communities. *Contrast* bee, beta, Gosset

α-epistemology: cuckoo statistical examples of, 203–6; origins in Karl Pearson and Ronald Fisher; representative examples of, 194–95, 200–202, 222–26. *For legacy of, see* copyright; standard error; statistical significance. *Contrast* β-epistemology; β-self

Altman, Douglas G.: measures misuse of statistical significance in medicine, 162, 163, 164; sarcastic formatting of *p* values, 62, 163

Altman, Morris: correlates standard error with social psychology of fear, 251; editorial declaration against misuse of statistical significance in *Journal of Socio-Economics,* 168; inspired book, xix

An *F*-test

"floccinaucinihilipilification /. . . / *noun.* [from Latin *flocci, nauci, nihili, pili,* words denoting 'at little value' + -fication.] The action or habit of estimating as worthless (*Shorter Oxford English Dictionary,* 5th ed., 985).

For example:

Ronald A. Fisher would say, "The potash manures are not statistically significant. Disregard them."

William S. Gosset would say, "Statistical significance is a *floccinaucinihilipilification,* a finding of little scientific value. If you want to know about the potash manures you have to consider their pecuniary value compared to the barley you're trying to make money with."